Springer Complexity

Springer Complexity is an interdisciplinary program publishing the best research and academic-level teaching on both fundamental and applied aspects of complex systems – cutting across all traditional disciplines of the natural and life sciences, engineering, economics, medicine, neuroscience, social and computer science.

Complex Systems are systems that comprise many interacting parts with the ability to generate a new quality of macroscopic collective behavior the manifestations of which are the spontaneous formation of distinctive temporal, spatial or functional structures. Models of such systems can be successfully mapped onto quite diverse "real-life" situations like the climate, the coherent emission of light from lasers, chemical reaction-diffusion systems, biological cellular networks, the dynamics of stock markets and of the internet, earthquake statistics and prediction, freeway traffic, the human brain, or the formation of opinions in social systems, to name just some of the popular applications.

Although their scope and methodologies overlap somewhat, one can distinguish the following main concepts and tools: self-organization, nonlinear dynamics, synergetics, turbulence, dynamical systems, catastrophes, instabilities, stochastic processes, chaos, graphs and networks, cellular automata, adaptive systems, genetic algorithms and computational intelligence.

The two major book publication platforms of the Springer Complexity program are the monograph series "Understanding Complex Systems" focusing on the various applications of complexity, and the "Springer Series in Synergetics", which is devoted to the quantitative theoretical and methodological foundations. In addition to the books in these two core series, the program also incorporates individual titles ranging from textbooks to major reference works.

Understanding Complex Systems

Founding Editor: J.A. Scott Kelso

Future scientific and technological developments in many fields will necessarily depend upon coming to grips with complex systems. Such systems are complex in both their composition – typically many different kinds of components interacting simultaneously and nonlinearly with each other and their environments on multiple levels – and in the rich diversity of behavior of which they are capable.

The Springer Series in Understanding Complex Systems series (UCS) promotes new strategies and paradigms for understanding and realizing applications of complex systems research in a wide variety of fields and endeavors. UCS is explicitly transdisciplinary. It has three main goals: First, to elaborate the concepts, methods and tools of complex systems at all levels of description and in all scientific fields, especially newly emerging areas within the life, social, behavioral, economic, neuro- and cognitive sciences (and derivatives thereof); second, to encourage novel applications of these ideas in various fields of engineering and computation such as robotics, nano-technology and informatics; third, to provide a single forum within which commonalities and differences in the workings of complex systems may be discerned, hence leading to deeper insight and understanding.

UCS will publish monographs, lecture notes and selected edited contributions aimed at communicating new findings to a large multidisciplinary audience.

For further volumes:
http://www.springer.com/series/5394

Guillaume Deffuant • Nigel Gilbert

Viability and Resilience of Complex Systems

Concepts, Methods and Case Studies from Ecology and Society

Springer

Editors

Dr. Guillaume Deffuant
Cemagref
avenue des Landais 24
63172 Aubière CX 1
France
guillaume.deffuant@cemagref.fr

Prof. Nigel Gilbert
University of Surrey
Centre for Research in Social Simulation
Guildford, GU2 7XH Surrey
United Kingdom
N.Gilbert@surrey.ac.uk

ISSN 1860-0832 e-ISSN 1860-0840
ISBN 978-3-642-20422-7 e-ISBN 978-3-642-20423-4
DOI 10.1007/978-3-642-20423-4
Springer Heidelberg Dordrecht London New York

Library of Congress Control Number: 2011934027

Cover design: SPi Publisher Services

Printed on acid-free paper

Springer is part of Springer Science+Business Media (www.springer.com)

Preface

This book is based on the research carried out during the PATRES (Pattern resilience) project, supported by the European Commission as part of its 6th Framework Programme. The project involved five European research teams, specialising in methods (computer science, applied mathematics, the physics of complex systems), and in the broad areas of application of the research: ecology and sociology.

The central concept of the book is resilience. This concept is having an impact in a wide variety of fields, such as ecology, sociology and psychology. Some people consider it to be key for designing and implementing a truly sustainable future. The basic idea behind resilience is the ability of a system to recover after strong perturbations. But when it comes to the details, there is much discussion among scientists and specialists. Mathematical definitions have the advantage of being precise and unambiguous, but they are criticised as too narrow. Fuzzier, more verbal definitions have a richness of meaning and may play a role as boundary concepts between different disciplines, facilitating exchange between them but they are not directly operational.

Our first motivation is to contribute to this debate. The initial challenge of the project was to build on a recent formalisation of the concept of resilience, based on viability theory (Martin 2004). This definition is founded on a precisely defined mathematical theory and is, in our view, closer to the intuitive concept than other existing mathematical definitions of resilience. Moreover, the viability based definition is oriented towards action on the system: it allows one to compute laws of actions on the system in order to keep or restore a desired property, lost after a perturbation. General algorithms for doing such computations exist.

However, solving a viability problem is in practice possible only when the problem is expressed in a state space of relatively small dimensionality (up to 7 or 8 dimensions). It is therefore impossible to apply the method to systems described by a large number of interconnected entities, because the state space has too many dimensions. Nevertheless, when the interconnected entities generate statistical regularities or patterns that can be described in a reasonable number of dimensions, and when the desired properties of the system are related to these patterns, the approach can be used. The association of patterns with resilience justified the title of the project.

The main objective of the project was derived from this scientific challenge: to elaborate efficient methods and tools for modelling and managing pattern resilience in complex systems. The methods integrate contributions from the research on resilience, more particularly its link with viability theory, and methods for pattern identification in models and data. The main objective therefore had two aspects:

- Defining more powerful and more flexible methods and tools for solving viability problems, using recent statistical tools such as Support Vector Machines (SVMs), in order to increase the range of systems for which the resilience problem can be solved.
- Providing a set of methods and tools for modelling pattern dynamics, building on current work on the exploration of models with systematic experimental designs, and on general statistical physics approaches. These methods and tools were tested on case studies, drawn from very different domains in ecology and the social sciences.

In the first part of the book, we introduce the concepts of resilience and viability.

The first chapter is a short introduction to the literature about resilience, concluding with the main challenges faced at the state of the art. We discuss the necessity of coining a definition that is neutral about the properties that the system should maintain. Choosing such properties implies that we have a priori values about what the system should or should not do. The choice of these values belongs to morals, politics, or business, not to science, although as scientists we can make an important contribution by identifying what options are available and whether any of them imply unrealisable outcomes.

The second chapter presents the viability based definition of resilience using an example, and compares it with more usual definitions. This chapter also introduces the main concepts of viability theory: the viability kernel and capture basin. It shows that the viability based definition is a generalisation of the so-called 'engineering' definition of resilience and also of other definitions based on attraction domains.

Then, in the second part, we test the approach on a set of case studies, which all start with an individual based model, defined by a population of interacting agents. Such dynamics have far too many variables to be compatible with the current tools for computing viability kernels and capture basins. Therefore, an important question addressed in the book is how to describe a complex dynamics using only a small set of synthetic variables.

The first case study is devoted to a model of competition between languages that is, for some values of its parameters, equivalent to a well known model of behaviour propagation, the voter model. This model is used to illustrate different techniques coming from physics for deriving synthetic dynamics from the individual interactions, and we show how viability based resilience can be computed for the case of the mean-field approach. We suppose that it is desirable to retain at least a minimum number of speakers of all the existing languages, and that an institution has some means of modifying the prestige of the languages.

In the case study on collaborative Web communities, we start by reviewing conceptual and empirical issues in the identification of desirable viable states for these systems. We present a summary of empirical analyses of their dynamics, focussing on two paradigmatic cases: peer production systems and social media groups. A simple model is then proposed in which the viability of these systems is assessed against constraints on group population and group content size. We discuss the interaction of these constraints, the autonomous demographic dynamics of such systems and the possible control actions which may be adopted to ensure or restore their viability.

In the savanna case study, we consider first a quite detailed model of the ecological dynamics, including a large number of variables, distributed over a set of spatial components. In this case, we first had to derive a simpler individual based model to be able to obtain a sound synthetic dynamics. It was necessary in this case to keep some indicators of the spatial pattern, using pair correlation in addition to mean field equations. After having described this simpler model, we consider the problem of maintaining the savanna through actions modifying the grazing level.

In the bacteria biofilm case study, we also start with a complex individual based model that we first simplify. In this case study, we again use a pair approximation approach to get synthetic information about the spatial pattern. However, instead of considering this information at two specific distances, as in the savanna case study, we derive a more global indicator for the shape of the pair approximation function. Then, the resilience problem that we address is to keep a particular spatial organization of the biofilm, together with bounds on the density of bacteria.

In the third part of the book, we describe in more detail tools and techniques used in the case studies.

We first describe Kaviar, a software prototype that computes viability kernels and capture basins that was developed during the project. It uses a learning technique known as support vector machines (SVMs). The chapter explains how learning techniques in general, and SVMs in particular, can be used to facilitate the computation of viability kernels and capture basins. The chapter also provides practical examples that are complementary to the software user guide.

The final chapter of the book describes how viability and resilience can be used to derive more robust policies of action. The idea is to compute the distance to the boundary of the viability kernel, and to define the action that keeps the system as far as possible from this boundary.

The research reported in this book has been carried out with the support of the European Commission through a Sixth Framework Programme European project (NEST-Complexity, number: 043268, acronym: PATRES) for which we are grateful.

Guillaume Deffuant
Nigel Gilbert

Contents

Contributors

Isabelle Alvarez Cemagref - LISC, 24 av. des Landais 63172 Aubière, France, isabelle.alvarez@cemagref.fr

Justin M. Calabrese Helmholtz Centre for Environmental Research (UFZ), Permoserstrasse 15, 04318 Leipzig, Germany; Smithsonian Conservation Biology Institute, National Zoological Park, 1500 Remount Rd., Front Royal, VA 22630, USA, CalabreseJ@si.edu

Xavier Castelló Institute for Cross-Disciplinary Physics and Complex Systems IFISC (CSIC-UIB), Campus Universitat Illes Balears, Palma de Mallorca, Spain, xavi@ifisc.uib-csic.es

Laetitia Chapel Cemagref - LISC, 24 av. des Landais, 63172 Aubière, France, laetitia.chapel@cemagref.fr; Université de Bretagne Sud, Lab-STICC 8 rue Montaigne, 56017 Vannes Cedex, France, laetitia_chapel@yahoo.fr

Guillaume Deffuant Cemagref - LISC, 24 av. des Landais, 63172 Aubière, France, guillaume.deffuant@cemagref.fr

Victor Eguìluz Institute for Cross-Disciplinary Physics and Complex Systems IFISC (CSIC-UIB), Campus Universitat Illes Balears, Palma de Mallorca, Spain, victor@ifisc.uib.es

Nigel Gilbert University of Surrey, Guildford, Surrey GU2 7XH, UK, N.Gilbert@surrey.ac.uk

Volker Grimm Helmholtz Centre for Environmental Research (UFZ), Permoserstrasse 15, 04318 Leipzig, Germany, volker.grimm@ufz.de

Lucia Loureiro-Porto Department of Spanish, Modern Languages and Latin, Universitat Illes Balears, Palma de Mallorca, Spain, lucia.loureiro@uib.es

Nabil Mabrouk Cemagref - LISC, 24 av. des Landais, 63172 Aubière, France, nabil.mabrouk@cemagref.fr

Sophie Martin Cemagref - LISC, 24 av. des Landais, 63172 Aubière, France, sophie.martin@cemagref.fr

Jean-Denis Mathias Cemagref - LISC, 24 av. des Landais, 63172 Aubière, France, jean-denis.mathias@cemagref.fr

Maxi San Miguel Institute for Cross-Disciplinary Physics and Complex Systems IFISC (CSIC-UIB), Campus Universitat Illes Balears, Palma de Mallorca, Spain, maxi@ifisc.uib-csic.es

Camille Roth University of Surrey, Guildford, Surrey GU2 7XH, UK; Centre d'analyse et de mathématique sociales, CNRS-EHESS, 54, boulevard Raspail, 75006 Paris, France, roth@ehess.fr

Dario Taraborelli Centre for Research in Social Simulation, University of Surrey, Guildford, Surrey GU2 7XH, UK, d.taraborelli@surrey.ac.uk

Federico Vasquez Institute for Cross-Disciplinary Physics and Complex Systems IFISC (CSIC-UIB), Campus Universitat Illes Balears, Palma de Mallorca, Spain, federico@pks.mpg.de

Part I
Concepts

Chapter 1
What Is Resilience? A Short Introduction

Volker Grimm and Justin M. Calabrese

1.1 Introduction

Agent-based complex systems such as economies, ecosystems, or societies, consist of autonomous agents such as organisms, humans, companies, or institutions that pursue their own objectives and interact with each other and their environment (Grimm et al., 2005). Fundamental questions about such systems address their stability properties: How long will these systems exist? How much do their characteristic features vary over time? Are they sensitive to disturbances? If so, will they recover to their original state, and if so, why, from what set of states, and how fast? These questions are so important because the mere existence of agent-based complex systems is, in contrast to many systems studied in physics or chemistry, not granted but intriguing, calling for an explanation (Jax et al. (1998)). The building blocks of these systems – organisms or human actors – do not have a blueprint of the entire system in mind and behave accordingly, but follow their own objectives. Nevertheless, system-level properties emerge which allow the identification of the systems and their behaviours over time. Tropical forests, for example, can be self-similar over thousands of years and reliably provide functions and services that are important for us. Systems can, however, also collapse and lose their identity and functions. For example, a stock market can crash, or a savanna can turn into

V. Grimm (✉)
Helmholtz Centre for Environmental Research-UFZ, Permoserstrasse 15, 04318 Leipzig, Germany
e-mail: volker.grimm@ufz.de

J.M. Calabrese
Helmholtz Centre for Environmental Research-UFZ, Permoserstrasse 15, 04318 Leipzig, Germany

Smithsonian Conservation Biology Institute, National Zoological Park, 1500 Remount Rd., Front Royal, VA 22630 USA
e-mail: CalabreseJ@si.edu

G. Deffuant and N. Gilbert (eds.), *Viability and Resilience of Complex Systems*, Understanding Complex Systems, DOI 10.1007/978-3-642-20423-4_1,

a scrubland due to overgrazing, rendering it useless as rangeland (Scheffer et al., 2009).

Understanding stability properties is thus not only of scientific interest but is also a prerequisite for successful management of agent-based complex systems. For example, how can ecosystems be used in a sustainable way without impairing their stability properties and therefore their potential to provide services also in the future? How should economies be regulated to prevent crashes? How should large companies organize their workflow so that it is not only efficient but also robust to disturbances?

Of all scientific domains dealing with agent-based complex systems, ecology seems to be the one where stability properties have been discussed and explored most intensively. Models have played a central role in the development of and debate surrounding stability concepts in ecology. Other disciplines have been so far less influenced by modelling, such as political science, or are dominated by equilibrium-centred approaches, such as economics, so that questions of why and how long systems exist have lower priority.

We will therefore in this chapter review stability concepts in ecology. In particular, we will focus on resilience, a concept that has recently been promoted by the Resilience Alliance, an international multidisciplinary network to improve the sustainable management of socio-ecological systems (Folke 2006; Brand 2005; for an overview of the use of the concept of resilience in socio-ecology, see Janssen et al. 2006; Janssen 2007; Walker et al. 2006). First we will give an overview of definitions and terms used in ecology. Then, we will focus on two different notions of resilience, 'engineering' and 'ecological' resilience and describe further central concepts promoted by the Resilience Alliance.

Although this chapter is based on approaches and examples from ecology and socio-ecology, it nevertheless addresses stability concepts and resilience in general, as emergent properties of agent-based complex systems. The purpose of this chapter is to introduce important stability concepts, provide verbal definitions and examples, and serve as a guide to the relevant literature. Chapter 2 will provide more specific mathematical definitions of resilience.

1.2 Stability Concepts in Ecology

In a literature review, Grimm and Wissel (1997) evaluated 163 definitions of 70 different stability terms from ecology, but even at that time more definitions and terms certainly existed. In the meantime, many new definitions have been added, in particular definitions of 'resilience' (Brand and Jax 2007).

Grimm and Wissel (1997) found that despite this terminological diversity, only three fundamentally different stability properties exist: constancy, resilience, and persistence (Table 1.1). All existing definitions of stability properties can be mapped to one of these basic properties or to a combination of them. Grimm and Wissel

(1997) concluded that it would not be appropriate to equate just one of these properties with 'stability'. Rather, 'stability' is a multi-layered concept that includes the three basic 'stability properties' as specific aspects. Three further concepts are important enough to be considered essential stability properties, but are related to the basic properties: resistance (an interpretation of constancy), elasticity, and domain of attraction (quantitative aspects of resilience). If there are only so few basic stability properties, why does this huge diversity of terms and definitions exist? Grimm and Wissel (1997) discuss three possible reasons.

First, the term 'stability' is ambiguous by itself and cannot be narrowed down to one of the properties in Table 1.1. 'Stability' is a concept comprising the different aspects listed in Table 1.1. Many researchers therefore add an adjective to 'stability', for example 'species deletion stability' (Pimm 1980, p. 142), to make the term more specific. Alternatively, they may use a narrower definition, for example equating 'stability' with 'resilience' as defined in Table 1.1, or they might simply invent a new term such as 'amplitude' (Connell and Sousa 1983, p. 790).

Second, the fascination with 'stability' reflects the desire of ecology for powerful concepts: "Stability belongs to the expressions (as information and energy) of which, sometimes, a global explanatory power is expected and which is supposed to make tedious attention to detail more or less superfluous." (Schwegler 1985, p. 263; translated from German).

Third, stability concepts can, with the exception of 'persistence', not be applied to entire systems but only to specific state variables characterizing these systems, for example total biomass, number of species, fixation of CO_2, or spatial patterns. Moreover, statements about stability properties also depend on the specific type of

Table 1.1 Six basic stability concepts, identified in a literature review in ecology (after Grimm and Wissel 1997)

Stability	Concept definition	Comment
Constancy	Staying essentially unchanged	Often, the inverse property 'variability' is considered
Resilience	Returning to the reference state (or dynamics) after a temporary disturbance	In theoretical ecology, this property often simply was referred to as 'stability'
Persistence	Persistence through time of an ecological system	This concept refers to entire systems, whereas the other concepts refer to one or more specific state variables
Resistance	Staying essentially unchanged despite the presence of disturbances	This is an interpretation of 'constancy'
Elasticity	Speed of return to the reference state (or dynamics)	This has often been referred to as 'resilience'
Domain of attraction	The whole of states from which the reference state (or dynamics) can be reached after a temporary disturbance	Related to 'persistence' since the 'domain' defines the states a system can achieve without losing its identity

disturbance considered, on the temporal and spatial scales involved, and on how, precisely, the reference state or dynamics is defined. The diversity of stability terms and definitions may thus reflect the many different ways in which ecosystems can be characterized and disturbed. Despite the profusion of terms in the ecological stability discussion, which can be confusing and irritating, two concepts play a dominant role in ecology: engineering and ecological resilience.

1.3 Engineeering Resilience

'Engineering resilience' is the same as 'elasticity' as defined in Table 1.1: "Rate and speed of return to preexisting conditions after disturbance" (Holling and Gunderson 2002). The qualifier 'engineering' was added to distinguish this notion of resilience from the more holistic notion of Holling (1973) (see next section). But why 'engineering'? Because it can easily be calculated from simple dynamical models representing communities of interacting populations (Otto and Day 2007). Prototypes of such models are Lotka–Volterra predator-prey or competitive systems. These models are expressed as nonlinear ordinary differential equations where the state variables are the time-dependent densities of the species considered (Wissel 1989).

Calculating elasticity, and checking for resilience, is straightforward using linear stability analysis: calculate equilibrium densities; apply infinitesimally small displacements, or disturbances, from the equilibrium; use Taylor expansions to obtain a set of linear equations describing the dynamics of the displacements; calculate the eigenvalues of the matrix of coefficients of the linearized set of equations. If the real part of the dominant eigenvalue is smaller than zero, the disturbed system will return to its equilibrium (e.g., Otto and Day 2007). Thus, the sign of the eigenvalue's real part indicates whether the system is resilient, and the inverse of its absolute value is a measure of its elasticity, or engineering resilience. However, the inverse of the eigenvalue indicates the time the system will need to return to equilibrium. A system might therefore be resilient in principle, but return so slowly that on time scales of practical relevance it is not.

For the 'domain of attraction', which is the second aspect of resilience as defined in Table 1.1, no similarly straightforward approach to calculate this property exists. Therefore, theoretical ecology had a much stronger focus on equilibria and return times than on the domain of attraction.

Theoretical ecologists were intrigued by being able to 'calculate' the 'stability' (as they usually called it) of ecological communities, which for the first time facilitated the quantitative study of one of the most important questions of ecology: the relationship between the diversity (and complexity) of a community and its stability (May 1974). Linear stability analysis and the concept of 'engineering resilience' were therefore dominating approaches in theoretical ecology in the 1980s and a large part of the 1990s.

However, many ecologists felt that these approaches reflected a quite narrow notion of the stability properties of ecosystems (Holling and Gunderson 2002). Specifically, the engineering resilience perspective does not allow the study of entire systems and how their internal organization and mechanisms promote persistence despite disturbances which could cause ecosystems to lose their characteristic features and functions. These researchers often referred to the highly influential review of Holling (1973) which suggested a more holistic definition.

1.4 Ecological Resilience

Holling's definition is verbal and qualitative, not mathematical and quantitative: "resilience determines the persistence of relationships within a system and is a measure of the ability of these systems to absorb changes of state variables, driving variables, and parameters, and still persist. In this definition resilience is the property of the system and persistence or probability of extinction is the result." (Holling 1973, p. 17). He also suggested two measures of resilience: "Since resilience is concerned with probabilities of extinction, firstly, the overall area of the domain of attraction will in part determine whether chance shifts in state variables will move trajectories outside the domain. Secondly, the height of the lowest point of the basin of attraction (...) above equilibrium will be a measure of how much the forces have to be changed before all trajectories move to extinction of one or more of the state variables." (p. 20). This definition, which later was slightly modified and sometimes termed 'ecological' or 'ecosystem's resilience' (Holling and Gunderson 2002; Brand 2005; Brand and Jax 2007), is, like 'stability', a term comprising several stability properties (sensu Table 1.1) simultaneously: persistence, resistance, resilience, and domain of attraction. Though Holling's review is widely cited, it has not been adopted by theoretical ecologists and modellers, because there is no simple way to quantify ecological resilience. Nevertheless, in the middle of the 1990s, a group of ecologists and social scientists founded the Resilience Alliance (www.resalliance.org). The Resilience Alliance has the declared aim to promote and develop Holling's notion of ecological resilience and related concepts because they are considered essential for solving vital socio-ecological problems and for fostering sustainability. The resilience definition currently preferred by the Resilience Alliance was formulated by Walker et al. (2004): "Resilience is the capacity of a system to absorb disturbance and reorganize while undergoing change so as to still retain essentially the same function, structure, identity, and feedbacks." The three main differences between ecological and engineering resilience are: (1) A shift in focus from equilibria to domains of attraction, often also called 'regimes', where a certain characteristic network, or regime, of processes controls the system's properties and functions. This is important because ecosystems usually are not in equilibrium but can change within relatively wide margins, without losing their identity. Savannas, for example, are driven by rainfall events. After a couple of years with higher rainfall, tree density can increase and tree distribution can become clustered (Jeltsch et al. 1999). However, trees and grasses still coexist, as required by the definition of savannas (Jeltsch et al. 2000). (2) A shift in focus from numerical values of state variables to 'relationships', i.e. to the internal organization of ecosystems which gives rise to their properties. (3) A shift in focus from the ability to recover after disturbance (engineering resilience) to the ability to 'absorb' the effect of disturbances, i.e. not to change essentially in the first place. Mechanisms are believed to be in place which buffer the effect of disturbances, as in the case of savannas. The most important implication of this is that this buffering ability can be lost, leading to an abrupt regime shift.

1.5 Regime Shifts

If environmental conditions change too much, for example due to climate change, human impact, or both, ecosystems can suddenly change to another regime which might no longer provide services essential for human well-being. This is in analogy to chemical buffers, which have only a certain capacity for buffering pH values.

A classical example involves shallow lakes (Scheffer and Carpenter 2003), which can tolerate increasing input of phosphorus only up to a certain tipping point where lakes turn from a clear to a turbid state. Figure 1.1 shows another example from semiarid savannas (Jeltsch et al. 1997): if the density of livestock exceeds a certain threshold, grass cover is reduced so much that the internal organization of savannas, which includes interaction between grass cover, fires, and tree density (Calabrese et al. 2010), is disrupted. As a consequence, woody cover is no longer controlled by fire and increases abruptly, leading to a state of this ecosystem that no longer can be used for livestock grazing. This new regime, sometimes also called an 'alternative state', is also resilient so that the loss of ecosystem function caused by the regime shift is irreversible, at least on time scales relevant to humans. Such transitions occur in savannas worldwide and are an alarming example of the unsustainable management of natural resources.

Regime shifts have been demonstrated for several ecosystems, in particular shallow lakes, savannas, and coral reefs. The Resilience Alliance maintains on their website a database of observed regime shifts. However, the question of whether all ecosystems show abrupt changes and are characterized by alternative states is still open (Schroeder et al. 2005). Nevertheless, regime shifts are the most important element of the resilience approach fostered by the Resilience Alliance. They make us focus on the risk of losing ecosystem functions that are vital to human well-being. Consequently, management should not be concerned about equilibria and some kind of 'balance of nature', but should instead focus on the key mechanisms that allow a system to persist, and on the fact that these mechanisms have only a certain capacity, which can be reduced by environmental change and human impact.

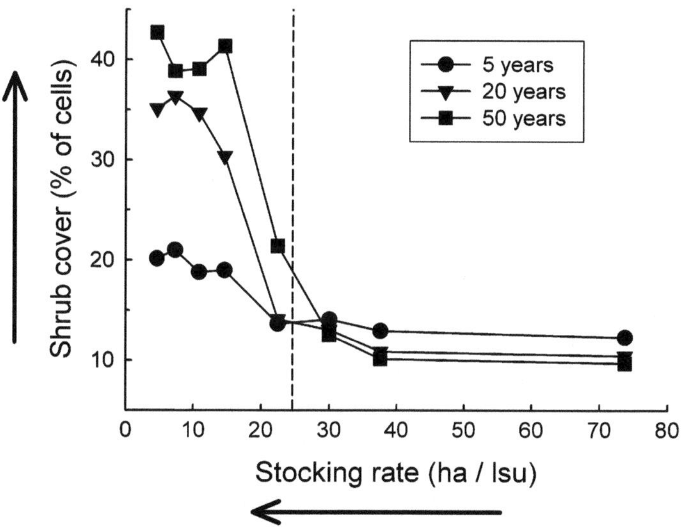

Fig. 1.1 Regime shift in a semiarid savanna, as predicted by a simulation model. If livestock density increases (i.e., stocking rate, measured as hectare per livestock unit [lsu], decreases) beyond a certain threshold, shrub cover increases abruptly, making the savanna unsuitable as rangeland. The change in shrub cover is not very marked after only 5 years of overgrazing. The *dashed vertical line* gives the estimated stocking capacity of the site, which was estimated empirically and independently of the model. (Redrawn after Jeltsch et al. 1997)

1.6 Adaptive Cycle and Panarchy

The conceptual framework of the Resilience Alliance includes two further main concepts: adaptive cycles and panarchy. Adaptive cycles are an attempt to provide the generic mechanism underlying resilience. Ecosystems are believed to be resilient because they are able to 'adapt' to changes and new conditions. Resilience is believed to be based on cyclic changes of two properties: potential and connectedness: "Potential sets limits to what is possible – it determines the number of the alternative options for the future. Connectedness determines the degree to which a system can control its own destiny (...). Resilience determines how vulnerable the system is to unexpected disturbances and surprises that can exceed or break that control" (Holling and Gunderson 2002, p. 51).

Connectedness is assumed to increase over time, leading to high internal control and limited potential to cope with disturbances. Naturally, such over-connected systems crash into a release period, where it has the potential to reorganize, thereby coping with disturbances. This development is believed to be cyclic. Resilience is also assumed to change at the local scale in a cyclic way. When connectedness is low, resilience is high because the system can vary over a wide range of states and respond to disturbance in many different ways. When connectedness, however, is high, ecosystem resilience is low because the system is more tightly organized and has fewer options for responding to disturbances. Interestingly, engineering resilience can be high at the same time as ecosystem resilience is high, i.e. effects of not too extreme disturbances quickly disappear (e.g. Holling 2001).

Because of its very general nature, the concept of the adaptive cycle should be considered a metaphor (Carpenter et al. 2001) or thinking tool rather than a testable scientific theory. This metaphor certainly contains important elements of what drives agent-based complex systems. For example, succession of different plant communities on a certain site includes many elements of the adaptive cycle. In particular, the so-called climax community, which could be a mature, old-growth forest, could be considered over-connected and prone to crashing following disturbances such as fire or pest outbreaks. However, succession and, similarly, adaptive cycles apply to smaller spatial units, not the entire ecosystem.

In the concept of 'panarchy', adaptive cycles on different temporal and spatial scales are coupled in a nested hierarchy (Holling and Gunderson 2002). Ecological and socio-ecological systems are thus assumed to be driven by 'cross-scale interactions' (Walker et al. 2004). As a result of these interactions, the characteristic control, or regime, of such systems emerges (Holling et al. 2002): "The complexity of adaptive systems can be traced to interactions among three to five sets of variables, each operating at a qualitatively distinct speed and scale." (Brand 2005).

1.7 Challenges of the Resilience Approach

The Resilience Alliance has been extremely prolific, producing hundreds of publications (Janssen et al. 2006; Janssen 2007) and a large number of books, and keeping a well-maintained website that contains databases, references, and material for education and policy makers. These activities have had a tremendous impact on how ecologists and social scientists think about socio-ecological systems, their stability properties, and sustainable management.

Nevertheless, proceeding from metaphors and thinking tools to operational concepts is challenging. Three main challenges are:

- *Separating normative from descriptive definitions of resilience.* This point has been raised by Brand and Jax (2007). They are concerned about the trend to mix descriptive definitions, which refer to how system are, with normative definitions, which refer to how systems should be. For example, Folke (2006) defines resilience as "The underlying capacity of an ecosystem to maintain desired ecosystem services in the face of a fluctuating environment and human use." (p. 14). Normative definitions have their place, for example in facilitating "communication across disciplines and between science and practice" (Brand and Jax 2007), but for operationalizing the concept of resilience clear descriptive definitions are needed.
- *Gaining mechanistic understanding.* The observations on which the resilience approach builds are certainly essential and contain information about how

agent-based complex systems are organized. For example, a central idea underlying the concept of panarchy is that such systems are usually controlled by a small number of, say, three to five variables. But why is this so? Understanding of the generative mechanisms (Lawson 1989) is key to putting concepts into practice and to successful management. For example, Thulke and Grimm (2010) show how computational models helped to devise successful strategies for controlling wildlife diseases.

- *Reconciling engineering and ecological resilience.* The defining feature of engineering resilience is that a mathematical protocol exists to calculate, for models formulated as differential equations and for small disturbances, whether or not a system returns to equilibrium and how fast return will be. Ecosystem resilience is more focussed on the domain of attraction and regime shifts. Mathematical approaches for dealing with these aspects of resilience exist (Anderies et al. 2002) but are less general and powerful than linear stability analysis. Their limitations will be discussed in the next chapter.

1.8 Summary and Conclusions

Stability is a multi-layered concept comprising the three elements: constancy, (engineering) resilience, and persistence (Table 1.1). For a long time, theoretical ecology focussed on one specific aspect of stability: whether or not a certain state variable returns to its reference value after a temporary disturbance. Linear stability analysis allowed to quantify this 'engineering' notion of resilience but its ecological relevance remained unclear. In contrast, the concept of 'ecological resilience', which is promoted by the Resilience Alliance, is by definition multi-layered and comprises the aspects of persistence, resistance, resilience, and domain of attraction (Table 1.1). This certainly is an achievement, because it allowed to shift the focus from equilibria to the functioning of ecosystems and the key question under what conditions agent-based complex systems lose their ability to cope with disturbances and environmental changes, which leads to regime shifts.

So far, however, the concept of ecological engineering has not been operationalized: it remains unclear how to quantify resilience and identify the mechanisms underlying resilience. In this book, the main approach to achieve operationalization of ecological resilience is: simplify and aggregate simulation models so that key mechanisms of resilience are easier to identify, and, if possible, apply viability theory and a related new concept of resilience (Martin 2004, Chap. 3). The approach studied in this book – viability theory – can be seen as an attempt to overcome some of the challenges of the resilience approach. A major aim of this book is augmenting the approach of engineering resilience, without losing its mathematical background, and linking it with a variety of complex dynamics defined by individual interactions.

In general, it will be important to clearly separate between *analytical* and *synthetic aspects* of resilience. Analytical stability concepts like constancy, resistance, and engineering resilience focus on single state variables and their dynamics. They are diagnostic tools for exploring how well different state variables capture the organization of a system, how different disturbances, reference states, and observation at different scales and hierarchical levels help understanding the functioning of agent-based complex systems. In contrast, synthetic concepts, such as persistence and ecological resilience, aim at explaining, in a holistic way, the existence and functioning of agent-based complex systems. They refer to the phenomenon we want to understand, explain, and take into account in managing such systems: their ability to cope with disturbance and change, and the limits of this ability.

Acknowledgements We would like to thank G. Deffuant, N. Gilbert, S. Martin, C. Roth, and D. Taraborelli for valuable comments on earlier drafts of this chapter.

References

Anderies JM, Janssen MA, Walker BH (2002) Grazing management, resilience, and the dynamics of a fire-driven rangeland system. Ecosystems 5:23–44

Brand FS (2005) Ecological resilience and its relevance within a theory of sustainable development. Technical report, UFZ

Brand FS, Jax K (2007) Focusing the meaning(s) of resilience: Resilience as a descriptive concept and a boundary object. Ecol Soc 12(1):23. [online] http://www.ecologyandsociety.org/vol12/iss1/art23/

Calabrese JM, Vazquez F, SanMiguel M, Lopez C, Grimm V (2010) The individual and interactive effects of tree-tree establishment competition and fire on savanna structure and dynamics. Am Nat 175:E44–E65

Carpenter SR, Walker BH, Anderies JM, Abel N (2001) From metaphor to measurement: resilience of what to what? Ecosystems 4:765–781

Connell JH, Sousa WP (1983) On the evidence needed to judge ecological stability or persistence. Am Nat 121:789–824

Folke C (2006) The economic perspective: Conservation against development versus conservation for development. Conserv Biol 20:686–688

Grimm V, Revilla E, Berger U, Jeltsch F, Mooij WM, Railsback SF, Thulk HH, Weiner J, Wiegand T, DeAngelis DL (2005) Pattern-oriented modeling of agent-based complex systems: Lessons from ecology. Science 310:987–991

Grimm V, Wissel C (1997) Babel, or the ecological stability discussions: An inventory and analysis of terminology and a guide for avoiding confusion. Oecologia 109:323–334

Holling CS (2001) Understanding the complexity of economic, ecological, and social systems. Ecosystems 4:390–405

Holling CS, Gunderson LH, Peterson GD (2002) Sustainability and panarchies. In: Gunderson LH, Holling CS (eds) Panarchy: Understanding transformations in human and natural systems. Island Press, Washington

Holling CS (1973) Resilience and stability of ecological systems. Annu Rev Ecol Evol Syst 4:1–24

Holling CS, Gunderson LH (eds) (2002) Resilience and adaptive cycles. In: Panarchy: Understanding transformations in human and natural system, Island Press, Washington

Janssen MA, Schoon ML, Ke W, Borner K (2006) Scholarly networks on resilience, vulnerability and adaptation within the human dimensions of global environmental change. Global Environ Change 16:240–252

Janssen MA (2007) An update on the scholarly networks on resilience, vulnerability, and adaptation within the human dimensions of global environmental change. Ecol Soc 12:9

Jax K, Jones GG, Pickett STA (1998) The self-identity of ecological units. Oikos 82:253–264

Jeltsch F, Milton SJ, Dean WRJ, vanRooyen N (1997) Analysing shrub encroachment in the Southern Kalahari: A grid-based modelling approach. J Appl Ecol 34:1497–1509

Jeltsch F, Moloney KA, Milton SJ (1999) Detecting process from snap-shot pattern: Lessons from tree spacing in the southern Kalahari. Oikos 85:451–467

Jeltsch F, Weber GE, Grimm V (2000) Ecological buffering mechanisms in savannas: A unifying theory of long-term tree-grass coexistence. Plant Ecol 150:59–78

Lawson T (1989) Abstraction, tendencies and stylised facts: A realist approach to economic analysis. Camb J Econ 13:59–78

Martin S (2004) The cost of restoration as a way of defining resilience: A viability approach applied to a model of lake eutrophication. Ecol Soc 9(2):8. [online] http://www.ecologyandsociety.org/vol9/iss2/art8/

May RM (1974) Stability and Complexity in Model Ecosystems. Princeton University Press, Princeton

Otto SP, Day T (2007) Mathematical Modeling in Ecology and Biology. Princeton University Press, Princeton

Pimm SL (1980) Food web design and the effect of species deletion. Oikos 35:139–149

Scheffer M, Bascompte J, Brock WA, Carpenter V, Brovkinand SR, Dako V, Held H, vanNes EH, Rietkerk M, Sugihara G (2009) Early-warning signals for critical transitions. Nature 461:53–59

Scheffer M, Carpenter SR (2003) Catastrophic regime shifts in ecosystems: Linking theory to observation. Trends Ecol Evol 18:648–656

Schroeder A, Persson L, de Roos AM (2005) Direct experimental evidence for alternative stable states: A review. Oikos 110:3–19

Schwegler H (1985) Ökologische Stäbilitat. Verh Ges Okol 13:263–270

Thulke HH, Grimm V (2010) Ecological models supporting management of wildlife diseases. In: Thorbek P, Forbes V, Heimbach F, Hommen U, Thulke HH, van den Brink PJ, Wogram J, Grimm V (eds) Ecological Models for Regulatory Risk Assessments of Pesticides: Developing a strategy for the future. Society of Environmental and Chemistry (SETAC) and CRC Press, Pensacola, pp 67–76

Walker B, Holling CS, Carpenter SR, Kinzig A (2004) Resilience, adaptability and transformability in social-ecological systems. Ecol Soc 9(2):5. [online] http://www.ecologyandsociety.org/vol9/iss2/art5

Walker B, Gunderson L, Kinzig A, Folke C, Carpenter S, Schultz L (2006) A handful of heuristics and some propositions for understanding resilience in social-ecological systems. Ecol Soc 11:13. [online] http://www.ecologyandsociety.org/vol11/iss1/art13/

Wissel C (1989) Theoretische Okologie. Springer, Berlin

Chapter 2
Defining Resilience Mathematically: From Attractors To Viability

Sophie Martin, Guillaume Deffuant, and Justin M. Calabrese

2.1 Introduction

The previous chapter presents different views of resilience, starting from Holling's conceptual definition of "ecological resilience": the capacity of a system to absorb "disturbance and reorganize while undergoing change so as to still retain essentially the same function, structure, identity, and feedbacks" (Walker et al. 2004). In this chapter, we focus on operational, mathematically precise definitions of resilience. In the literature, the main mathematical definitions of resilience are based on dynamical systems theory, and more specifically on attractors and attraction basins (also related to 'regime shifts' presented in the previous chapter). We present these definitions in detail, and illustrate their utility on a relatively simple rangeland management model. Furthermore, we use the rangeland example to highlight some key limitations of attractor based definitions of resilience.

We then introduce the viability based definition (Martin 2004). In this approach, the desired property (function or service) of the system is defined as a given subset of the state space (containing attractors or not). This is more general than the attractor based definition in which the desired property is defined as a subset of attractors.

We underline that changing this definition of the desired properties of the system changes the mathematical framework for defining resilience, and that viability theory is the appropriate one (Aubin 1991). Indeed, viability theory focuses on

S. Martin · G. Deffuant (✉)
Cemagref - LISC, 24 av. des Landais 63172 Aubière, France
e-mail: sophie.martin@cemagref.fr; guillaume.deffuant@cemagref.fr

J.M. Calabrese
Helmholtz Centre for Environmental Research-UFZ, Permoserstrasse 15, 04318 Leipzig, Germany

Smithsonian Conservation Biology Institute, National Zoological Park, 1500 Remount Rd., Front Royal, VA 22630 USA
e-mail: CalabreseJ@si.edu

G. Deffuant and N. Gilbert (eds.), *Viability and Resilience of Complex Systems*, Understanding Complex Systems, DOI 10.1007/978-3-642-20423-4_2,

dynamical systems which cease to function or badly deteriorate when they cross the limits of a subset of their state space, called the viability constraint set. Hence the problem addressed is to keep the system within the limits of this viability constraint set. The problem is formally the same, when the system should remain in a desired set. The concepts and tools from viability theory can be used directly. Viability theory methods and tools have recently been used to address some of the issues encountered in natural resource management (Béné et al. 2000, Mullon et al. 2004, Doyen et al. 2007).

We use two main concepts derived from viability theory: "viability kernel" and "capture basin". Within this framework, one can define the resilience basin as the *capture basin of the viability kernel.* We present these concepts in more detail, and illustrate them on an example of savanna simplified dynamics (Anderies et al. 2002). We show that this viability based definition of resilience includes the attractor based definition as a particular case. Moreover, it offers new possibilities:

- The approach allows one to compute policies of action to keep or restore the desired property of the system
- The desired property can be defined as a subset of the state space which does not include any attractor of the dynamical system. In this case, to keep the desired property, one should act regularly on the system. This corresponds to usual situations of ecological or social systems, which are impossible to address in the attractor based framework of resilience

2.2 Attractor Based Definition of Resilience

In this section, we summarise the main mathematical definitions of resilience based on attractors and attraction basins.

2.2.1 Main Hypothesis: The Dynamics Includes "Good" and "Bad" Attractors

First, we quickly introduce the notions of dynamical systems and attractors (for mathematical details, see for instance Murray et al. 1994).

A dynamical system is defined by several state variables $x \in \mathbb{R}^n$, and an equation stating how these variables evolve with time. Generally, this equation gives the value of the derivative of $x \in \mathbb{R}^n$ with time as a function of x.

$$\begin{cases} x'(t) = f(x(t)), \\ x(t_0) = x_0. \end{cases} \tag{2.1}$$

For the computer scientist who will discretise the time into small steps of size dt, this equation gives the value of $x(t+dt)$ as a function of $x(t)$. A first order approximation is:

$$\begin{cases} x(t+dt) = x(t) + f(x(t)).dt, \\ x(t_0) \quad = \quad x_0. \end{cases} \tag{2.2}$$

Starting from point $x_0 \in \mathbb{R}^n$, one can therefore derive a trajectory which associates with time t the value $x(t)$ of the state variables at time t.

An equilibrium point $x_0 \in R^n$ is a point where the dynamics is stopped ($f(x_0) = 0$). This equilibrium is asymptotically stable if it tends to attract the states in its vicinity. It is unstable if some points in its vicinity tend to be rejected by it. Figure 2.1 represents, for a one-dimensional state variable, a stable and an unstable equilibrium as a ball in a landscape. On the peaks, the ball can be at equilibrium, but a small change in its position leads it to fall to one side or the other. On the contrary, the ball returns to the bottom of a valley when it is slightly displaced from it.

If $x_0 \in \mathbb{R}^n$ is an equilibrium, we call attraction basin the set gathering all points $x \in \mathbb{R}^n$ such that $\lim_{t\to+\infty} ||x(t) - x_0|| = 0$. Figure 2.2 displays an illustration of three stable equilibria with their attraction basins.

More generally, an attractor is a set of states (points in the state space), invariant under the dynamics, towards which neighbouring states in a given basin of attraction asymptotically approach in the course of dynamic evolution. An attractor is defined as the smallest unit which cannot be itself decomposed into two or more attractors with distinct basins of attraction. Asymptotically stable equilibria are attractors, such as limit cycles. There are also strange attractors, but we will not discuss them in this chapter.

The attractor based definition of resilience supposes that the dynamics of the system is organized around such attractors and attraction basins, and that the system is generally close to an attractor, except immediately after strong perturbations that

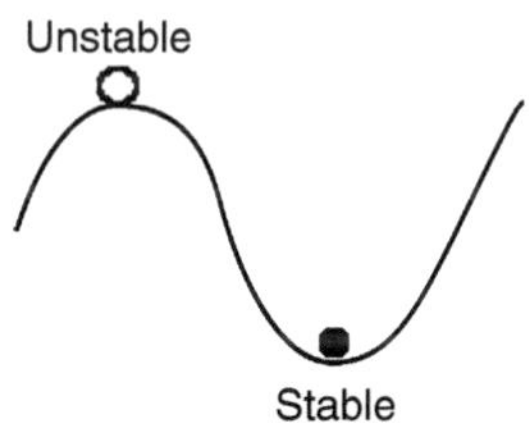

Fig. 2.1 Intuitive representation of a stable and an unstable equilibrium in a one-dimensional state space

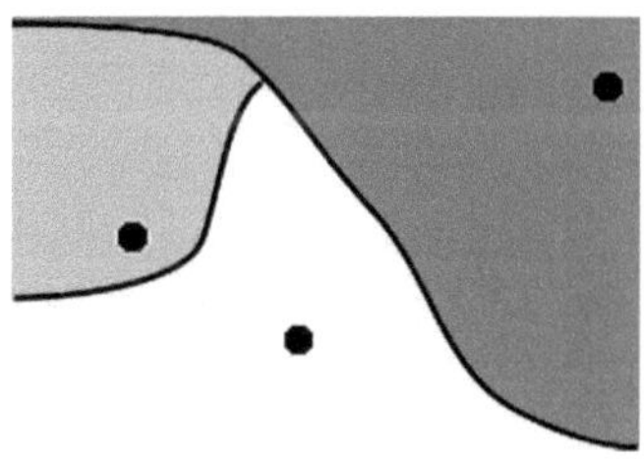

Fig. 2.2 Example of a two-dimensional state space with 3 stable equilibria and their attraction basins

can drive it far from its attractors. Moreover, it is supposed that some attractors correspond to desired (good) functioning of the system, whereas others correspond to situations that should be avoided.

2.2.2 *Illustration on Simplified Dynamics of Savanna*

To make these concepts more concrete, we illustrate them with a stylized model of rangeland management (Anderies et al. 2002). The model explores the effects of physical, ecological, and economic factors on the resilience of a rangeland system. A key problem in rangeland systems is to balance the conflicting objectives of increasing short-term profit while preserving the long-term sustainability of the rangeland. Increasing the stocking rate (animals/ha) of the rangeland increases short-term gain, but can lead to effectively irreversible degradation of the rangeland if the resultant grazing pressure is too high. Alternatively, keeping the stocking rate too low can lead to substantial loss of income. The goal of rangeland management then is to identify policies of action (changes in the stocking rate) that best reconcile these opposing objectives.

In the model of Anderies et al. (2002), good stable equilibria correspond to high shoot biomass and the bad ones to null level of shoot biomass (Fig. 2.3).

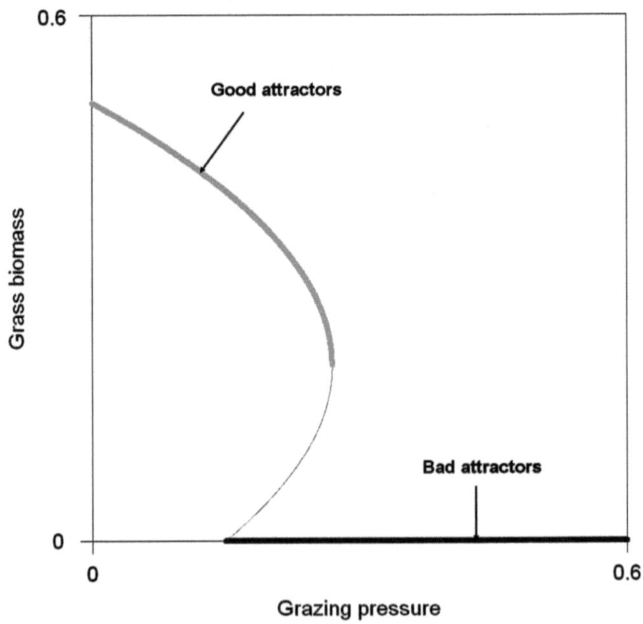

Fig. 2.3 Equilibria of the rangeland system model (2.3). Stable equilibria (attractors) are drawn with *plain lines* and unstable ones are drawn with *dashed lines*. 'Good' attractors (high shoot biomass) are coloured *grey* and 'bad' ones (null shoot biomass) are coloured *black*

After some simplifications in order to make the presentation easier, the evolution of the shoot biomass (grass) can be considered to be governed by the differential equation:

$$s'(t) = s(\alpha_1 + \alpha_2 s - \alpha_3 s^2) - \gamma_g s := f(s), \tag{2.3}$$

where α_1, α_2 and α_3 are parameters, and γ_g is the grazing pressure. The first term expresses the nonlinear growth of the grass (which is actually coupled with the growth of bushes but we simplified), and the second term expresses the effect of grazing on the grass biomass. The equilibria are given by:

$$\begin{cases} s \geq 0 \\ s = 0 \\ \alpha_1 - \gamma_g + \alpha_2 s - \alpha_3 s^2 = 0. \end{cases} \tag{2.4}$$

Taking the Anderies et al. values for the parameters, we obtain the sets of stable and unstable equilibria plotted in Fig. 2.3. Notice that $s = 0$ is unstable for grazing pressure below 0.15 and stable for grazing pressure strictly greater than 0.15. The equilibria defined by

$$0.15 - \gamma_g + 1.2s - 3s^2 = 0 \tag{2.5}$$

are stable for $s > 0.2$ and unstable for $s \in [0; 0.2]$.

The first step to evaluate resilience consists in distinguishing good from bad attractors according to the desired system state. In the rangeland model, "good" attractors are high shoot biomass stable equilibria and "bad" ones are null shoot biomass stable equilibria (see Fig. 2.3).

2.2.3 *Attractor Based Measures of Resilience*

As far as real dynamical system models are concerned, the stability analysis is used to characterize the system response to small perturbation qualitatively: does the system return to its original state after a small perturbation, or not?

Resilience is then considered as a quantitative additional characterization either linked to the return time or the attraction basin size in the state space of the "good" asymptotically stable equilibria.

2.2.3.1 Resilience As the Inverse of Return Time

The return time is often used as a resilience index in the literature. As far as differential equation models are concerned, Pimm and Lawton (1977) and

DeAngelis (1980) for instance, used the eigenvalue with the maximal real part of the linearization to evaluate the resilience as the asymptotic rate of convergence. Actually, the asymptotic rate of convergence equals the opposite of the eigenvalue with the maximal real part (see for instance Murray et al. (1994)), and the bigger the rate of convergence is the more resilient the system is, since its state goes back more quickly to a neighbourhood of the equilibrium. To complement this index, which refers to the properties of asymptotic dynamics, Neubert and Caswell (1997) proposed indices allowing the characterization of transient responses following perturbation.

In the case of individual-based models and cellular automata, resilience is studied with simulations as the inverse of the time needed after some kind of disturbance to return to its original state (Ortiz and Wolff 2002) or to reach a certain percentage of the previous abundance (Matsinos and Troumbis 2002).

Keeping to the example of the rangeland system model of Anderies et al. (2002), the dynamics in the neighbourhood of good attractor s^* can be linearly approximated:

$$f(s^* + \epsilon) \approx \frac{df}{ds}(s^*).\epsilon \tag{2.6}$$

$$\frac{df}{ds}(s) = \alpha_1 + 2\alpha_2 s - 3\alpha_3 s^2 - \gamma_g. \tag{2.7}$$

Taking the Anderies et al. values for the parameters, we obtain:

$$\frac{df}{ds}(s) = 0.15 + 2.4s - 9s^2 - \gamma_g. \tag{2.8}$$

Considering a good attractor $s^* > 0.2$, defined by (2.5) according to the grazing pressure γ_g, the value of the linearized dynamics at s^* equals :

$$\frac{df}{ds}(s^*) = s^*(1.2 - 6s^*). \tag{2.9}$$

The resilience value, $R(s^*)$, of the system at equilibrium s^* is then the opposite of $\frac{df}{ds}(s^*)$.

Actually, $s^* = 0.2 + \sqrt{0.09 - \frac{\gamma_g}{3}}$ so s^* is defined for $\gamma_g \in [0; 0.27]$. When $\gamma_g \leq 0.27$,

$$R(\gamma_g) = -s^*(1.2 - 6s^*). \tag{2.10}$$

When $\gamma_g >= 0.27$, the resilience $R(\gamma_g)$ is null as there is no good asymptotically stable equilibrium associated with such grazing pressure intensity.

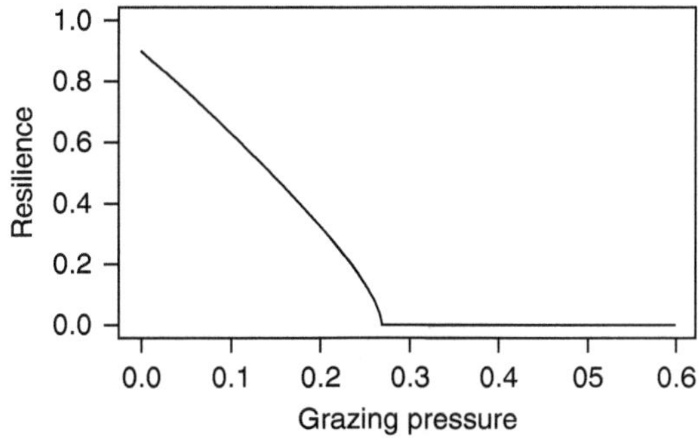

Fig. 2.4 Resilience value according to the grazing pressure γ_g, when defined as the asymptotic rate of convergence to a good attractor

Figure 2.4 shows the resilience value according to the grazing pressure γ_g. It is worth noting that this resilience measure refers to asymptotic dynamical properties, and is represented by the inverse of the asymptotic rate of convergence.

2.2.3.2 Resilience Proportional to the Attraction Basin Size

According to Beddington et al. (1976), resilience may be interpreted as the perturbation magnitude a system can absorb without experiencing qualitative changes. From a dynamical system viewpoint, "qualitative changes" may be considered as a state jump into another attraction basin. The resilience of the system is then defined as proportional to the distance between a good attractor and the boundary of its attraction basin (see for instance Collings and Wollkind 1990 and van Coller 1997).

In Anderies et al. (2002), this measure is used to study resilience of the rangeland model. To define a resilience measure based on the distance to the attraction basin boundary, we have to compare this distance with the distance to the definition domain (Fig. 2.5).

Variable s represents the shoot (grass) biomass and is therefore positive. When a good equilibrium exists, resilience is then defined as the quotient of the distance to the boundary of the attraction basin by the difference between this distance and the distance to the definition domain :

$$R(s^*) = \frac{\min\left(s^*, 2\sqrt{0.09 - \frac{\gamma_g}{3}}\right)}{s^* - \min\left(s^*, 2\sqrt{0.09 - \frac{\gamma_g}{3}}\right)} \tag{2.11}$$

Resilience is then infinite when the definition domain is included in the attraction basin of good attractors. When it is included in the attraction basin of bad attractors ($\gamma_g > 0.27$), resilience is null. Figure 2.6 displays the resilience plot.

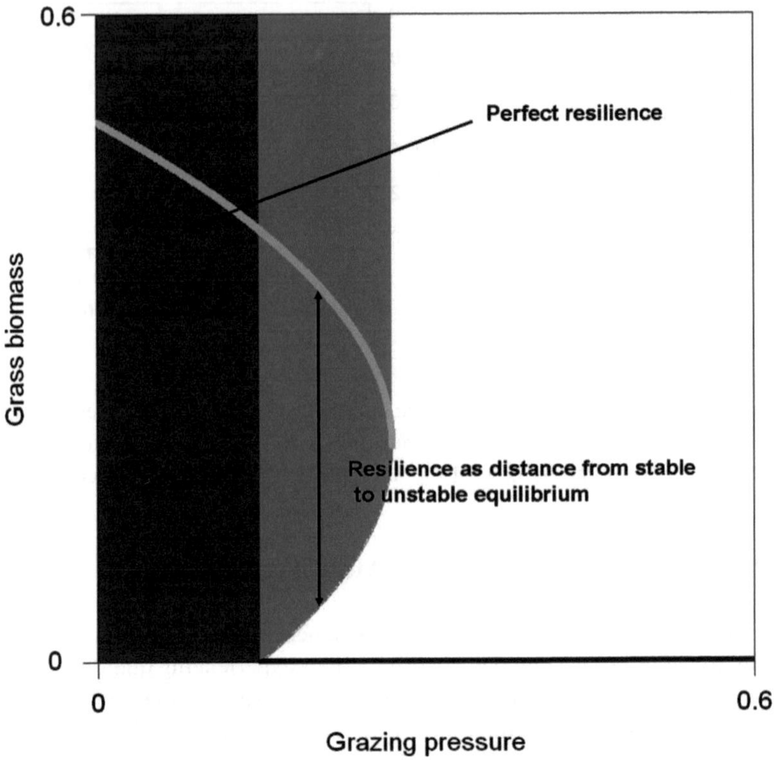

Fig. 2.5 Illustration of the distance between the good attractors and the boundary of their attraction basin

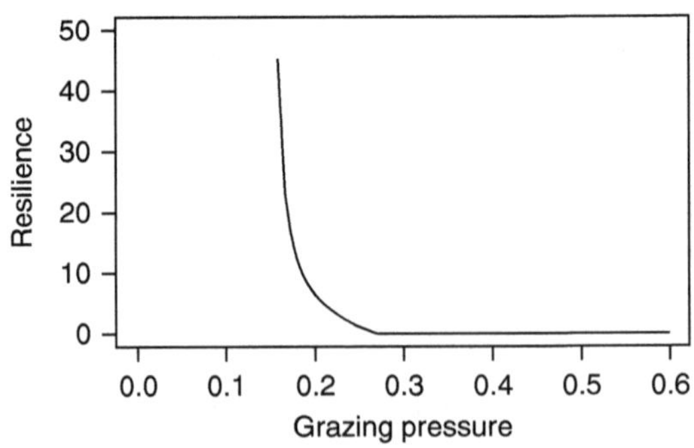

Fig. 2.6 Resilience measure based on the distance of the good attractors to the boundary of their attraction basin

It is worth noticing that such a definition of resilience is globally associated with an equilibrium. It does not distinguish between the states inside the attraction basin, whereas one could expect to get different resilience values for states near the attraction basin boundary and those which are closer to the equilibrium.

2.2.4 *Limitations*

2.2.4.1 The Choice of Actions is Not Included

These formalizations of resilience suppose that the feedback law or the management policy is defined, and the results only concern this particular management policy. Computing these resilience indices for different management policies allows one to compare them, but the set of possible policies is generally impossible to explore sufficiently to guarantee that a good policy is found. One needs another mathematical framework to include the choice of actions in the definition of resilience.

2.2.4.2 The Property of Interest Might not be Found in the Attractors

Both definitions suppose that the system shows the properties of interest when it reaches a given set of attractors (the good attractors). This corresponds to a widespread view in ecology that the desirable state of the system is an attractor where the system naturally goes without human intervention. But one can also often meet situations where human action and ecological dynamics are strongly interrelated. In this case, the property which we want to evaluate the resilience of is not necessarily found in an attractor.

Furthermore, inside a particular attraction basin, the resilience index should depend on the distance from the state of the system to the attraction basin boundary: resilience may be smaller near the boundary than in the middle of the attraction basin.

2.3 Viability Based Definition of Resilience: Dynamics without Management

In this section, we consider the definition of resilience based on the viability approach (Martin 2004) without including management actions. This will allow us to show that this definition can, in some conditions, coincide with the definition presented in the previous section. We shall demonstrate later that the viability based definition of resilience is more general, and can overcome some of the limitations of the attractor based definition.

2.3.1 *Resilience of a Property Defined by a Subset of the State Space*

In the attractor based definition of resilience, the desired property of the system is defined by a set of "good" equilibria or attractors. The main novelty of the viability

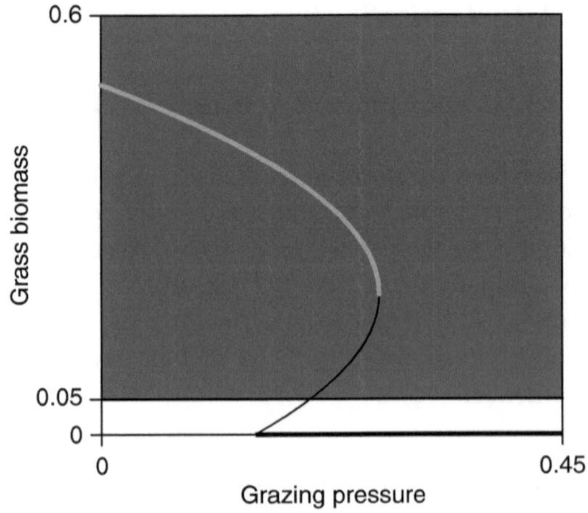

Fig. 2.7 The desired set, in *dark grey*, is defined as a level of grass higher than a threshold

based approach is to define the desired property of the system as a set of states, which are not necessarily attractors. This means that the desired property can be defined without knowing the dynamics of the system (its attractors for instance), but simply according to the desirable use of the system. For instance, in our simple rangeland model, a natural desired property of the system would be to have a grass biomass higher than a threshold (avoiding the complete disappearance of grass which would prevent the possibility of grazing). Therefore, the desired set can be, for example, as in Fig. 2.7.

This additional freedom in defining the desired property of the system has major consequences. First, we consider the resilience of the desired property rather than the resilience of the system itself. This is an important shift which requires us to be more precise. Moreover, since the considered states defining the desired property are not necessarily attractors, it is not possible any more to base the definition of resilience on attractor basins or convergence rates to the attractors. One needs a different theoretical framework, which focuses on the trajectories that remain in the desired set, or trajectories that come back to the desired set and remain in it. Fortunately, this theoretical framework exists: it is viability theory.

2.3.1.1 Viability Kernel

Viability theory was developed in the 1990s (Aubin 1991). Originally, its purpose was to study systems which collapse or badly deteriorate if they leave a given subset of the state space, called the viability constraint set. Therefore, the objective is to keep the system in the part of the state space where it can survive, i.e. where it is viable. The theory includes the case when one can act on the system to modify its trajectory, but for the moment, we focus on the particular case where no management is included (2.1), to facilitate comparison with the attractor based definition (we shall consider the case including management actions in the next section).

In this case, a viability problem is to identify the trajectories of a dynamical system which remain in the viability constraint set, and distinguish them from the trajectories that cross its boundaries. We recognise strong similarities with the resilience issue when the desired property of the system is defined as a set of states. The only difference is that the desired property of the system is defined by its usage, and can be very different from its viability. But these are simply different interpretations of the same mathematical concepts. In viability theory, the set of trajectories which remain in the constraint set is called the *viability kernel.* More formally, if we denote $S_f(x,t) \in \mathbb{R}^n$ the point reached at time t for a trajectory starting at point $x \in \mathbb{R}^n$ at time 0, then the viability kernel $Viab_f(K) \subset \mathbb{R}^n$ of desired set $K \subset \mathbb{R}^n$ for dynamics defined by (2.2) is defined by the following equation:

$$Viab_f(K) := \{x \in K \text{ such that } \forall t \geq 0, S_f(x,t) \in K\} \tag{2.12}$$

In the rangeland model example, with the desired set K defined previously, we can compute the associated viability kernel using general methods that are presented in more detail in Chap. 7 (Deffuant et al. 2007). The result is shown in Fig. 2.8. All the points leading to grass extinction are excluded, because they cross the threshold of grass biomass. Only the points of the initial desired set leading to the good attractors are kept in the viability kernel. Thus the viability kernel is the intersection between K and the attraction basin of the "good" attractors.

2.3.1.2 Resilience Basins

In a resilience problem, we want to address also the case where a perturbation leads the system to lose the desired property, and how the system can recover it. Hence

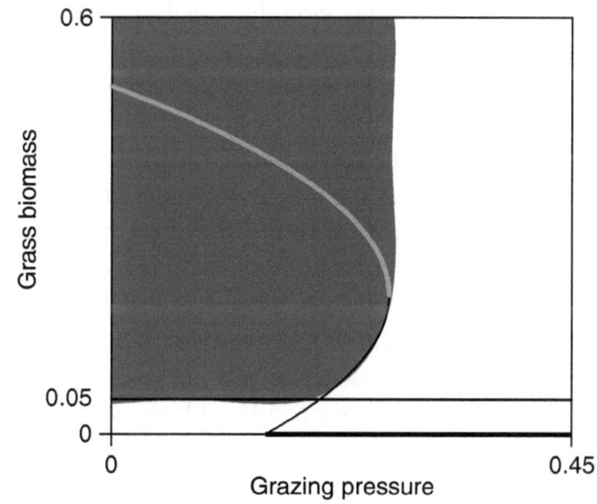

Fig. 2.8 The viability kernel in *dark grey* includes only the points from which the trajectory never crosses the limit of the constraint set

we need to get the states located outside K which go to K and remain in K. By definition the points which remain in K belong to the viability kernel, therefore, we need the points outside the kernel that go to the kernel. Actually, this problem to reach a given set as target is also addressed by viability theory (it can be seen as a particular case of the viability problem). The set of points going to a target set is called the *capture basin* of this set. More formally, if we consider a target set C, the capture basin $Capt_f(C)$ of C through dynamics defined by function f is given by:

$$Capt_f(C) = \{x \in \mathbb{R}^n \text{ such that } \exists T > 0 \text{ with } S_f(x,T) \in C\}. \tag{2.13}$$

In this framework, *the capture basin of the viability kernel defines the set of resilient states*. Indeed these are the states which go to the viability kernel of the desired property, and by definition, they remain there. In this framework, the viability kernel is analogous to the attractor in the usual framework, and the capture basin is analogous to the attraction basin. The set of resilient states is called the *resilience basin*.

Moreover, one can associate a resilience value with a point of the space as the inverse of the time to get to the viability kernel. The states belonging to the viability kernel have an infinite resilience because the time to reach the viability kernel is null. The states from which the viability kernel can be reached in finite time, have a finite strictly positive resilience value: they are resilient. Non-resilient states have a null resilience value: the time to reach the viability kernel is infinite. More generally, it is possible to define a cost for the restoration (return to the viability kernel) in this framework (Martin 2004). However, to simplify, in the following we shall suppose that the cost is only counted in time.

It is also convenient to define *finite time resilience basins*, which correspond to the states of the space which can be driven back to the viability kernel in a time which is lower than a given finite value. We shall see that practically we need to compute a set of such finite time resilience basins to derive the action policies to drive back the system into its viability kernel (see Chap. 7). In the following, we may refer simply to resilience basins, the plural implying that they correspond to finite time. Using the singular 'resilience basin' generally refers to the infinite time resilience basin.

Again, we can illustrate these concepts on our example of simplified savanna dynamics. Using the same general tools as before, one can compute resilience basins, which are represented by lines of different grey levels at the bottom of Fig. 2.9. Note that the return time gets higher and higher when the points get closer to the unstable equilibrium (boundary between attraction basins). It even reaches the limits of the computational scheme in the vicinity of the boundary between the attraction basins because there is a small strip which is considered as non-resilient, whereas theoretically, it is resilient. The dynamics is very slow in this part of the space, and the necessary time to leave it is approximated as being infinite.

If we compare this result with the attractor based definition of resilience, we note that most of the states which are resilient in the attractor based definition are viable in the viability based definition. The reason is that in the viability based

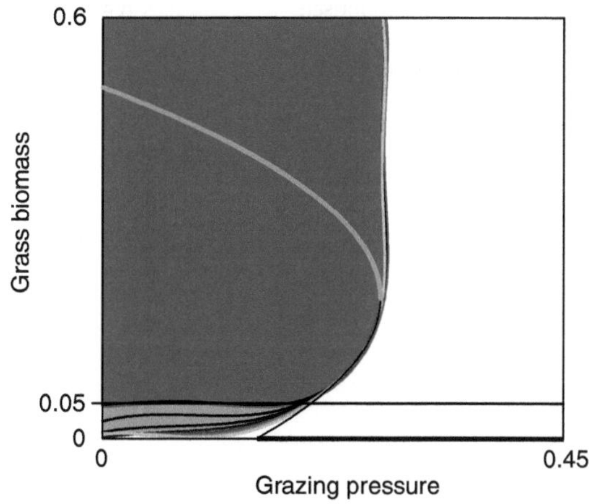

Fig. 2.9 The resilience basins include the viability kernel and some states below the constraint. The *lines* represent the resilience basins limits, and they get lighter when the resilience decreases. The *white area* is outside the resilience basins and corresponds to a null resilience (impossible to come back to the viability kernel)

definition, we only require the grass biomass to be higher than a threshold. All states satisfying this constraint are considered equivalent. Does it mean that the definitions are fundamentally different? Actually no, it depends on the definition of the desired state, as shown in the next section.

2.3.2 *Attractor Based Resilience as Particular Case of Viability Based Resilience*

In the attractor based resilience the desired set is the set of good attractors: it is assumed that the system shows the desired properties when it reaches one of these good attractors. Hence, it is interesting to test the viability based approach in the case where the desired set is the set of good attractors.

Actually, because of computational constraints for computing viability kernels, it is difficult to consider a desired set with a null surface. We therefore defined a set that surrounds the set of good attractors, as shown in Fig. 2.10.

The viability kernel of K gathers all states in K from which there exists an evolution that remains in K. Consequently, the viability kernel of K for the dynamics described by (2.3), is the intersection between K and the attraction basin (Fig. 2.11).

Having determined the viability kernel of the constraint set, we address the issue of determining whether points lying outside this viability kernel can enter it. The light grey part of the graph in Fig. 2.12 represents the infinite time resilience basin and the level curves represent growing time resilience basins (lighter curves indicate higher time).

Notice that in this case, when K is defined as a neighbourhood of the good attractors, the resilience basin and the attraction basin coincide. Therefore, both approaches lead to similar conclusions; the resilient states are the same. When

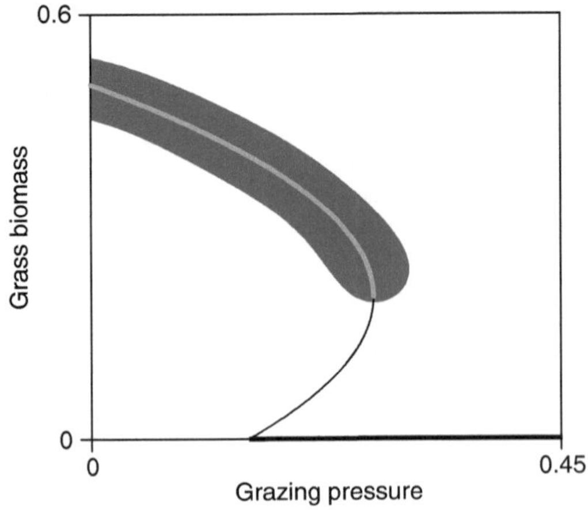

Fig. 2.10 The constraint set, K, coloured *dark grey*, is a neighbourhood of the good attractors

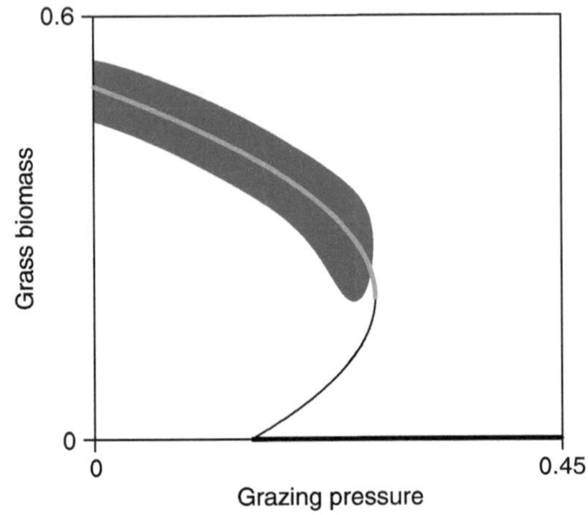

Fig. 2.11 The viability kernel of K is coloured *dark grey*. Only a small part of the constraint set is removed

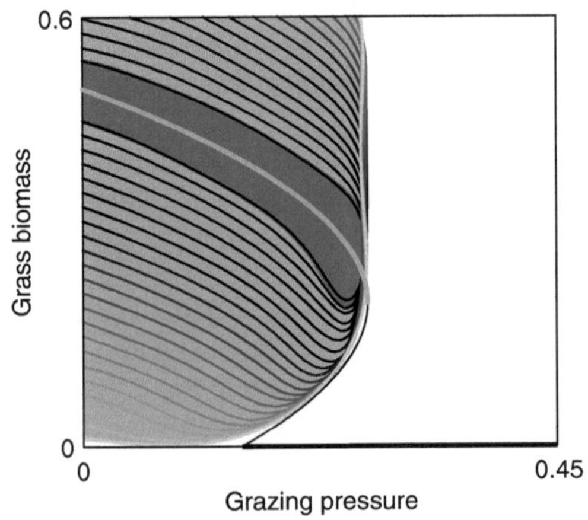

Fig. 2.12 The *dark grey* area is the viability kernel of K which coincides with the infinite resilience. The *lines* represent the limits of resilience basins, getting lighter when the resilience decreases. In the *white area*, the resilience is null

the desired set is defined as the vicinity of a set of attractors, the resilience basin coincides with the union of attraction basins. In this respect, the attractor based definition appears as a particular case of the viability based definition.

However, the resilience values are different. In the viability framework, the velocity of the dynamics close to the boundaries of the attraction basin is taken into account. For instance, it can be seen that the dynamics is very slow in the vicinity of the unstable equilibria, leading to a resilience tending to 0 (Fig. 2.12).

2.4 Viability Based Definition of Resilience: Including Management Actions

We suppose now that the evolution of the system also depends at each time t on a management action $u(t)$, with $u \in U \subset \mathbb{R}^p$. The differential equation governing the evolution of the system becomes:

$$\begin{cases} x'(t) = f(x(t), u(t)), \\ x(t_0) = \quad x_0. \end{cases} \tag{2.14}$$

When the time is discretised into small steps of size dt, we get the first order approximation:

$$\begin{cases} x(t + dt) = x(t) + f(x(t), u(t)).dt, \\ x(t_0) \quad = \quad x_0. \end{cases} \tag{2.15}$$

As mentioned previously, one strong limitation of attractor based resilience is that it cannot be used with such a system. Indeed, this system may include an infinity of trajectories starting from each point of the state space, corresponding to the infinity of action policies. Hence the attractor based approach can be used only if a policy of action $u(.) : t \rightarrow u(t) \in \mathbb{R}^p$ is a priori defined. We show now that the viability based resilience can be used to determine action policies that preserve or restore the desired property of the system.

2.4.1 Management Policy to Keep the Desired Property

Viability theory in its complete form addresses the problem of determining action policies for keeping a system inside a constraint set K. The viability kernel that we already presented above is a central concept of the theory to determine such a policy. When management actions are included in the system, the viability kernel (Aubin 1991) gathers all states from which there exists at least one action policy that keeps the system indefinitely inside K. More formally, if we denote $S_f(x, t, u(.))$ as the state reached after time t, starting from state x and applying action policy $u(.)$, with the controlled dynamical system (2.14), the viability kernel is defined by (2.16).

$$Viab_{f,U}(K) := \{x \in K \text{ such that } \exists u(.) \text{ with } \forall t \geq 0 \; S_f(x,t,u(.)) \in K\} \tag{2.16}$$

For instance, in the rangeland model, we suppose that one can modify the grazing pressure, γ_g, which is the shoot biomass offtake per unit area. Previously, we considered that this pressure was constant. In practice, γ_g is almost certainly not constant. Furthermore, land managers may make decisions to modify this grazing pressure to control the rangeland system. In the simplified approach of this toy model, we suppose that the grazing pressure can be modified at each time t of a value $\gamma_g'(t)$, with

$$-0.02 \leq \gamma_g'(t) \leq 0.02.$$

Indeed, it makes sense to suppose that the grazing cannot increase or decrease indefinitely in a short time step.

The system dynamics is then described by the controlled dynamical system:

$$\begin{aligned} s'(t) &= s(\alpha_1 + \alpha_2 s - \alpha_3 s^2) - \gamma_g s \\ \gamma_g'(t) &= u \in [-0.02; 0.02]. \end{aligned} \tag{2.17}$$

If we keep the desired set around the good attractors (Fig. 2.10), we get a viability kernel which is almost the same as the desired set (Fig. 2.13) – only a very small part on the right has been erased. It is logical that this viability kernel is bigger than in the case of a constant grazing, because modifying grazing offers the possibility to change the trajectories, and make them remain in K.

The viability kernel is important because it can be directly used to determine the action policies that keep the system inside the desired set. One is the "lazy"

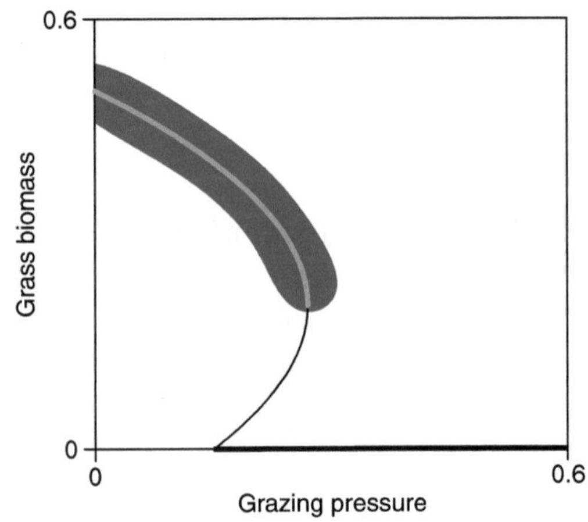

Fig. 2.13 The viability kernel of K almost coincides with the constraint set for the dynamics (2.17) including the possibility of modifying the grazing pressure with maximal rate equal to 0.02

policy, which requires that the set of actions U includes the possibility of doing nothing (action $u = 0$). Its principle is to do nothing while it is anticipated that the system does not cross the border of the viability kernel at the next time step. If it is anticipated that the system crosses this border, then is applied the first action that keeps the system inside it (one knows that such an action always exists, by definition of the viability kernel). More formally the procedure is as follows (in the case of discrete time):

- Start at point $x_0 \in \mathit{Viab}_{f,U}(K)$
- Denote $S_f(x_0, t, (u)_t)$ the point reached at time t, $(u)_t = (u_0, u_1, \dots, u_{t-1})$, being the list of actions applied at each of the t time steps. Suppose $S_f(x_0, t, (u)_t) \in \mathit{Viab}_{f,U}(K)$. Then:
  - If $S_f(x_0, t+1, (u)_{t+1}) \in \mathit{Viab}_{f,U}(K)$ with $u_t = 0$, then set $u_t = 0$
  - Otherwise, set $u_t = u$ with u such that

$$S_f(x_0, t+1, (u)_{t+1}) \in \mathit{Viab}_{f,U}(K).$$

  One can be sure that such an action u exists because $S_f(x_0, t, (u)_t) \in \mathit{Viab}_{f,U}(K)$. It can generally be found by a simple search, optimisation procedure or projection onto the viability kernel boundary (see Chaps. 7 and 8.1 for details)

2.4.2 *Management Policy to Drive the System Back to the Viability Kernel*

The original definition of a capture basin refers to controlled systems with management actions (2.16). The capture basin of target set C, denoted $\mathit{Capt}_{f,U}(C)$ is the set of states from which there exists a management policy leading the system into target set C. More formally, the definition of the capture basin is:

$$\begin{aligned} \mathit{Capt}_{f,U}(C) = \{ & x \in \mathbb{R}^n \text{ such that } \exists u(.) \\ & \exists t^* > 0 \text{ with } S_f(x, t^*, u(.)) \in C \}. \end{aligned} \tag{2.18}$$

The capture basin of the viability kernel defines also the set of resilient states (the resilience basin) in the case with management actions. Indeed, this set contains the states for which there exists a management policy driving back the system into the viability kernel, where it is then possible to keep the desired property indefinitely (in the absence of perturbations). The concept is the natural extension of the one presented in the case without management actions. The concept of finite time resilience basins also extends directly.

The property of interest cannot be restored from the states outside the resilience basin, and their resilience is therefore null. The states belonging to the resilience

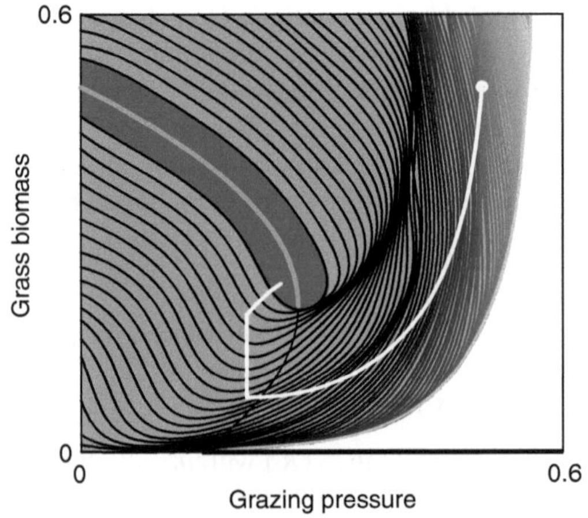

Fig. 2.14 Resilience basins for the property defined by K and dynamics described by (2.17) including the possibility of modifying the grazing pressure with maximal rate equal to 0.02. The *dark-grey area* is the viability kernel of K which coincides with the infinite resilience. The *grey lines* represent the limit of resilience basins, becoming lighter when the resilience decreases. The *white area* corresponds to a null resilience. The *white line* represents the trajectory of the system, under management actions derived from the resilience basins. It starts on the top right, outside the viability kernel, and the management actions on grazing drive back the system to the viability kernel

basin have a resilience value that is strictly positive and which may be quantified by the inverse of the time necessary to restore this property, or even with a more elaborate cost function.

As before, we can use general algorithms to apply the approach to our savanna model when management actions are considered. We note that the resilience basin is significantly larger than in Fig. 2.14, where the case without management actions was considered. This difference is expected because changing the grazing pressure provides new possibilities to drive back the system into the viability kernel.

Moreover, the computation of the set of capture basins at different time horizons (represented by the lines of different grey levels in Fig. 2.14), enables us to define an action policy that drives back the system to the viability kernel. The principle is to choose the action that makes the system go in the direction where the cost (time in the simplest case) decreases the most rapidly. In other words, this is the direction of the highest slope (considering the action possibilities) down to cost 0 which corresponds to the viability kernel. More details are provided about this in Chap. 7. Figure 2.14 gives an example of a trajectory computed using these management actions.

Therefore, we see here that the viability based definition of resilience is a generalization of the attractor based definition. It includes the attractor based definition as a particular case (when the desired set is a neighbourhood of a set of attractors and when there is no choice of action on the system), and allows us to

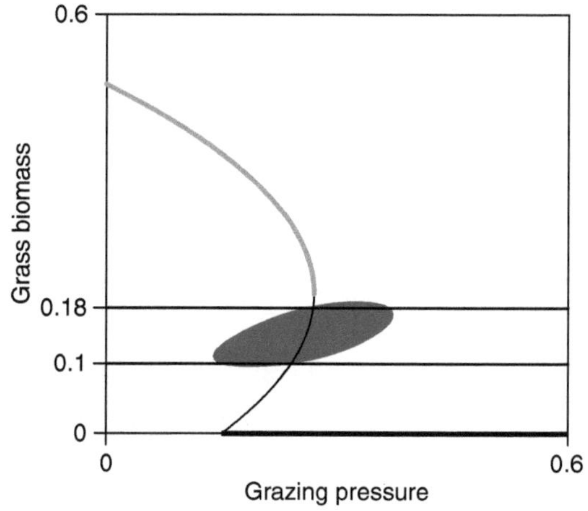

Fig. 2.15 The viability kernel of $K = [0; +\infty] \times [0.1; 0.18]$ for the dynamics (2.17) is coloured *dark grey*. K is *bordered black*

determine policies of action to keep the desired property of the system, or to recover it in the shortest possible time (or lowest cost).

2.4.3 Desired Set Without Attractor

Previously, we pointed out another limitation of the attractor based definition: the need to define the desired set as a set of attractors. The viability approach allows us to overcome this limitation. Indeed, it is possible to define a desired set without attractors, and still have a non-void viability kernel. In this case, however, it is necessary to act on the system regularly, to keep it within the desired set.

Our simplified model of savanna can illustrate this possibility. Suppose that we want to keep the level of shoot biomass between 0.1 and 0.18. This is a bit artificial, but one can imagine that the level of grass should not be too high because one wants to maximise the use of the resource. Note that this constraint set contains no attractor. We keep in the dynamics the ability of modifying the grazing pressure with absolute maximal rate 0.02 (2.17). The viability kernel is shown in Fig. 2.15.

We then compute the resilience basins, which are displayed in Fig. 2.16. The figure also shows one trajectory computed using the action policy derived from the resilience basins. As this example demonstrates, the viability framework can be applied when the constraint set does not include any attractor.

2.5 Conclusion and Perspectives

In this chapter, we adopted the idea that the desired property of a system can be defined as a subset of the state space without any specific conditions. This is a significant change compared with the usual mathematical approach, which supposes

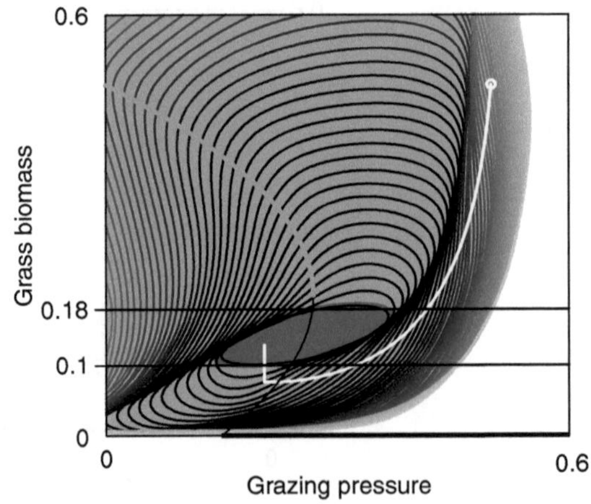

Fig. 2.16 Resilience basins for the property defined by $K = [0; +\infty] \times [0.1; 0.18]$ and dynamics described by (2.17) including the possibility of modifying the grazing pressure with maximal rate equal to 0.02. The *dark-grey area* is the viability kernel of K which coincides with the infinite resilience. The *grey lines* represent resilience basins, becoming lighter when the resilience decreases (and cost increases). In the *white area* the resilience is null. The *white line* represents the trajectory of the system under the management actions derived from the resilience basins. It starts outside the viability kernel, at the top right, and the management actions drive back the system to the viability kernel

that the desired property is defined by a set of attractors. Indeed, this leads to the adoption of viability theory as a mathematical framework for defining resilience. We can draw a parallel between attractor and viability based definitions of resilience:

- "Good" attractors are replaced by the viability kernel of the constraint set representing the desired property.
- The attraction basin of good attractors is replaced by the capture basin of the viability kernel (i.e. the points for which there exists a policy of action leading to the viability kernel).
- The resilience value as a convergence rate close to the good attractor (or as a measure of the size of the attraction basin), is replaced by the inverse of time (or more generally the cost) for driving back the system into the viability kernel.

The mathematical framework of viability theory leads to important conceptual and practical changes, compared with the attractor based framework:

- The resilience value depends on the state of the system and on its properties of interest which are defined by a subset of the state space, whereas in the attractor based framework, the resilience value is associated with an attraction basin.
- It is possible to derive action policies to keep or restore the desired property, whereas this is not the case in the attractor based framework.
- It is possible to define a desired set without any attractor, whereas this is impossible in the attractor based framework. It is worth noting that in the case

of an empty viability kernel, the capture basin of this viability kernel would also be empty, and hence there would be no resilient state, which is consistent, as no matter what the initial condition is, the desired property cannot be maintained.

We argue that the viability approach of resilience (Martin 2004) expresses better the original meaning of resilience which is to keep or restore a desired property of the system (Holling 1973). In our view, it gives a precise mathematical interpretation to this general concept, which generalises the current definitions based on attractors, without betraying the intuitive sense of the concept. One important asset of the approach is the possibility of using general algorithms to compute viability kernels, capture basins, and the associated policies of actions. However, these methods have some limitations, which are explained in more detail in Chap. 7.

The main limitation is that the computational complexity increases very rapidly with the dimensionality of the state space. Applying the approach directly on individual based models with dozens of state variables is totally excluded. Nevertheless, this method can still be used on these models, if it is possible to synthesise them into a simple dynamical system which represents adequately the main features of the dynamics.

In the next chapters of the book, we present several case studies where this global approach is applied: we consider a complex individual based model, then we synthesise it into a simplified dynamical model including a low number of state variables, and we use it to compute the viability and resilience of desirable properties.

References

Anderies JM, Janssen MA, Walker BH (2002) Grazing management, resilience, and the dynamics of a fire-driven rangeland system. Ecosystems 5:23–44

Aubin JP (1991) Viability Theory. Birkhauser, Basel

Beddington JR, Free CA, Lawton JH (1976) Concepts of stability and resilience in predator-prey models. J Anim Ecol 45:791–816

Béné D, Doyen L, Gabay D (2001) A viability analysis for a bio-economic model. Ecol Econ 36(3):385–396

Collings JB, Wollkind DJ (1990) A global analysis of a temperature-dependent model system for a mite predator-prey interaction. SIAM J Appl Math 50(5):1348–1372

DeAngelis DL (1980) Energy flow, nutrient cycling, and ecosystem resilience. Ecology 61(4): 764–771

Deffuant G, Chapel L, Martin S (2007) Approximating viability kernels with Support Vector Machines. IEEE Trans Automat Contr 52(5):933–937

Doyen L, DeLara M, Ferraris J, Pelletier D (2007) Sustainability of exploited marine ecosystems through protected areas : A viability model and a coral reef case study. Ecol Modell 208(2–4):353–366

Holling CS (1973) Resilience and stability of ecological systems. Annu Rev Ecol Syst 4:1–24

Martin S (2004) The cost of restoration as a way of defining resilience: a viability approach applied to a model of lake eutrophication. Ecol Soc 9(2):8. [online] http://www.ecologyandsociety.org/vol9/iss2/art8/

Matsinos YG, Troumbis AY (2002) Modeling competition, dispersal and effects of disturbance in the dynamics of a grassland community using a cellular automaton. Ecol Modell 149:71–83

Mullon C, Curry P, Shannon L (2004) Viability model of trophic interactions in marine ecosystems. Nat Resour Model 17:27–58

Murray RM, Sastry SS, Li Z (1994) A Mathematical Introduction to Robotic Manipulation. CRC Press, Inc., Boca Raton, FL, USA.

Neubert MG, Caswell H (1997) Alternatives to resilience for measuring the responses of ecological systems to perturbations. Ecology 78(3):653–665

Ortiz M, Wolff M (2002) Dynamical simulation of mass-balace trophic models for benthic communities of north-central Chile: assessment of resilience time under alternative management scenarios. Ecol Modell 148:277–291

Pimm SL, Lawton JH (1977) Number of trophic levels in ecological communities. Nature 268: 329–331

van Coller L (1997) Automated techniques for the qualitative analysis of ecological models : Continuous models. Conserv Ecol 1(1):5

Walker B, Holling CS, Carpenter SR, Kinzig A (2004) Resilience, adaptability and transformability in social-ecological systems. Ecol Soc 9(2): 5. [online] http://www.ecologyandsociety.org/vol9/iss2/art5/

Part II
Case Studies

Chapter 3
Viability and Resilience in the Dynamics of Language Competition

Xavier Castelló, Federico Vazquez, Víctor M. Eguíluz, Lucía Loureiro-Porto, Maxi San Miguel, Laetitia Chapel, and Guillaume Deffuant

3.1 Introduction

The study of language dynamics has been addressed from at least three different perspectives: language evolution (or how the structure of language evolves), language cognition (or the way in which the human brain processes linguistic knowledge), and language competition (or the dynamics of language use in multilingual communities). The latter is the approach followed in this chapter in which, therefore, we focus on problems of social interactions. We aim to contribute to the study of the complex phenomenon of language survival (viability), thoroughly studied in linguistics, from the perspective of pattern resilience.

The fact that 97% of the people in the world speak about 4% of the extant languages and that 50% of the 6,000 languages are expected to die in the current century (Crystal 2000) has made scholars struggle to identify and study the factors that may determine the survival or disappearance of most of our linguistic heritage as well as the mechanisms that could be implemented so as to revitalize an endangered

X. Castelló (✉) · V.M. Eguíluz · M. San Miguel
Institute for Cross-Disciplinary Physics and Complex Systems, IFISC (CSIC-UIB), Campus Universitat Illes Balears, E07122, Palma de Mallorca, Spain
e-mail: xavi@ifisc.uib-csic.es

F. Vazquez
MPI-PKS, Max Plank Institute for the Physics of Complex Systems, Nothnitzer Strasse 38, 01187 Dresden, Germany

L. Loureiro-Porto
Department of Spanish, Modern Languages and Latin, Universitat Illes Balears, E07122, Palma de Mallorca, Spain

L. Chapel
Université de Bretagne sud, Lab-STICC, 8 Rue Montaigne, 560I7 Vannes Cedex, France

G. Deffuant
Cemagref - LISC, 24 av. des Landais 63172 Aubière, France

G. Deffuant and N. Gilbert (eds.), *Viability and Resilience of Complex Systems*, Understanding Complex Systems, DOI 10.1007/978-3-642-20423-4_3,

language (cf., for example, Fishman 1991; Grenoble and Whaley 1998, 2006; Nettle and Romaine 2000; Hinton and Hale 2001; Bradley and Bradley 2002; UNESCO 2003; Wölck 2004; Tsunoda 2005). A motivation for many scholars is to understand how language endangerment occurs so as to avoid it or put it to an end. Such has been the aim of works such as Fishman's Reversing Language Shift (Fishman 1991), where he proposes two four-stage steps at each of which social and political actions should be taken so as to revitalize an endangered language. The first step consists of reversing language shift to attain diglossia and the second one addresses reversing language shift to transcend diglossia. Here we are concerned with this second step which is mathematically related to viability theory. Fishman (1991) has been the inspiration for a number of authors interested in language planning policies who, in the last decades, have tried to determine the degree of endangerment of a given language, as well as to measure the extent to which a given language may be revitalized stemming from any of these stages (see for example, Manley 2008, who assesses the role of micro-prestige among speakers of Quechua in Cuzco).

Language survival has also been subject of study of UNESCO (2003), who created an Ad Hoc Expert Group on Endangered Languages, so as to identify languages on a path toward extinction. The degree to which a language is actually bound to disappear may be assessed following different scales that account for each of the factors that have usually been identified as determinant in the vitality of a language. UNESCO (2003) has isolated nine of these factors (related to population, intergenerational transmission, and linguistic policies, among others). Although none of them can properly be considered in isolation to determine the viability of a language (they must be interconnected so as to thoroughly ascertain how endangered a given language is), in this work we focus on the role of two of such factors:

1. Governmental and institutional language attitudes and policies, including official status and use
2. Community members' attitudes toward their own language

Thus we have considered a class of models of socially interacting agents to describe language competition, featuring two parameters associated with these two factors. Factor (1) is taken into account by a parameter measuring the prestige of the language. In fact, the prestige of a language has been considered as one of the main factors affecting language competition since Labov's *Sociolinguistic Patterns* (Labov 1972). It measures the status associated to a language due to individual and social advantages related to the use of that language, being higher according to its presence in education, religion, administration and the media. Factor (2) is taken into account by a volatility parameter, a property which is not so often discussed in the linguistic literature as prestige is. These two parameters were already considered by Abrams and Strogatz (2003). For the parameter values that they explored in connection with Spanish-Quechua, Scottish-English and Welsh-English competitions, the prediction is that one of the languages eventually disappears. But reality provides myriads of counter-examples to this, in many cases achieved by active linguistic policies (e.g. French and Flemish in Belgium, Spanish and Catalan in Catalonia, etc.). We will describe combined ranges of prestige and volatility that make language coexistence viable.

We try to assess the viability and resilience of the language diversity in the line of Chap. 7 and Martin (2004). Resilience is seen as the capacity of a system to restore its properties of interest, which it may have lost after some perturbations. In language competition, these perturbations may be achieved by different situations, such as a military conquest (as was the case of Spanish in Latin America, in which the native languages were strongly threatened by the language of the conquerors), or massive immigration (e.g. the arrival of thousands of Spanish-speaking people to Catalan-speaking areas in Catalonia in the 1960s, as a result of industrial development). In our model, a language will be considered resilient if, after a perturbation like any of these, adequate political actions can restore its viability and, therefore, guarantee its survival.

This chapter intends to be a contribution to the understanding of the mechanisms underlying processes of social interaction at work in the dynamics of language competition, as well as the consequences of these mechanisms as regards language survival or extinction and the viability of language coexistence. We proceed in three steps. In the first step (Sect. 3.2) we introduce Individual Based Models (IBMs) of language competition, and explore, through computer simulations, the pattern dynamics of these models and the qualitative role of the prestige and volatility parameters. In the second step (Sect. 3.3) we discuss the derivation of macroscopic descriptions of these models, and we discuss how the macroscopic descriptions capture key aspects of the phenomena observed in the IBMs. Finally, in Sect. 3.4 we present an explicit calculation of viability and resilience based on a macroscopic description. A summary of conclusions is given in Sect. 3.5, while technical mathematical details of the micro–macro connection are contained in the Appendix.

3.2 IBMs for Language Competition

In this section we present two IBMs for language competition: the Abrams–Strogatz model (AS) and its extension allowing for bilingual agents, the Minett–Wang model (MW). After introducing the corresponding transition probabilities and the parameters of the model, we give a qualitative description of the role played by these parameters.

3.2.1 The Abrams–Strogatz model

The microscopic version (i.e. individual based) (Stauffer et al. 2007) of the AS-model (Abrams and Strogatz 2003) is a two-state model proposed for the competition between two languages. An agent i sits in a node within a social network of N individuals and has k_i neighbours. A neighbour means here another agent with which agent i has a social interaction. Agents can be in either of two states: X, agent using language X (monolingual X); or Y, agent using language Y (monolingual Y).

The state of an agent evolves according to the following dynamical rules: starting from a given initial condition, at each step we choose one agent i at random and we compute the local densities for each of the language states in the neighbourhood of node i, $\sigma_{i,l}$ (l=X, Y). The agent changes its state according to the following transition probabilities:

$$p_{i,X\to Y} = (1-S)(\sigma_{i,Y})^a \;, \qquad p_{i,Y\to X} = S(\sigma_{i,X})^a \tag{3.1}$$

Equation (3.1) give the probabilities for an agent i to change from community X to Y, or vice versa. They depend on the local densities ($\sigma_{i,X}$, $\sigma_{i,Y}$) and on two parameters: the *prestige* of the language X, $0 \leq S \leq 1$; and the *volatility*, $a \geq 0$. Prestige is a language property measuring the different status between the two languages, that is, the more prestigious language is the one which gives an agent more possibilities in the social and personal spheres. The case of socially equivalent languages corresponds to $S = 1/2$ (language X is more prestigious for $S > 1/2$). The volatility is a parameter characterizing social dynamics which gives shape to the functional form of the transition probabilities. The case a = 1 is the neutral situation of random imitation of a neighbour, where the transition probabilities depend linearly on the local densities. A high volatility regime exists for $a < 1$, with a probability of changing language state above the neutral case, and therefore agents change their state rather frequently. A low volatility regime exists for $a > 1$, with a probability of changing language state below the neutral case, where agents have a larger resistance to change their state. In this way, the volatility parameter gives a measure of the degree of accommodation or resistance of the agents to change their language use.

3.2.2 The Bilinguals Minett–Wang model

We consider here an extension of the AS-model proposed by Minett and Wang,[1] which takes into account the presence of a third possible state Z associated with bilingual agents using[2] both languages, X and Y. There are three local densities to compute for each node i: $\sigma_{i,l}$ ($l = X, Y, Z$). The agent changes its state according to the following transition probabilities:

---

[1]Notice that this extension was proposed in a working paper in 2005 (see also Wang and Minett 2005). The final version of the paper (Minett and Wang 2008) differs slightly on the transition probabilities. However, we analyse here their initial proposal.

[2]Notice that we consider *use* of a language rather than *competence*. In this way, learning processes are out of reach of the present model. Effectively, the situation is such as if all agents were competent in both languages.

$$p_{i,X\to Z} = (1-S)(\sigma_{i,Y})^a, \qquad p_{i,Y\to Z} = S(\sigma_{i,X})^a. \tag{3.2}$$

$$p_{i,Z\to Y} = (1-S)(1-\sigma_{i,X})^a, \qquad p_{i,Z\to X} = S(1-\sigma_{i,Y})^a. \tag{3.3}$$

Equation (3.2) give the probabilities for changing from a monolingual community, X or Y, to the bilingual community Z, while (3.3) give the probabilities for an agent to move from the Z community towards the X or Y communities. Notice that the latter depend on the local density of agents using the language to be adopted, including bilinguals. It is important to stress that a change from state X to state Y or vice versa, always implies an intermediate step through the Z-state ($p_{i,X\to Y} = p_{i,Y\to X} = 0$).

3.2.3 Qualitative Dynamics of IBMs

An implementation of these two IBMs in a two-dimensional regular network with four neighbours per node has been performed by designing a Java Applet in which one can tune the parameters described above, set different initial conditions, and see the simulations in real time.[3] The following descriptive overview of the models with different parameter settings gives insights on the emergent complex behaviour of these models, including issues of linguistic domain growth, linguistic boundaries, language coexistence, survival and extinction, and the role of bilingual agents.

- Neutral volatility. ($a = 1$)

  In the case of socially equivalent languages ($S = 0.5$), we observe in both models (AS and MW) a formation and growth of monolingual domains (see Fig. 3.1). However, the growth of these linguistic domains and the motion of linguistic boundaries has been shown to be due to different mechanisms: interfacial noise (AS-model) and curvature reduction (MW-model) (Castelló et al. 2006; Vazquez et al. 2010). Notice that the bilingual agents never form domains but, instead, they place themselves at the boundaries between monolingual ones. Finally, one of the two languages takes over the system. Due to the equivalent prestige, this happens for each of the languages with equal probability.

  The well known general role of prestige is clear when $S \neq 0.5$: the most prestigious language dominates, causing the extinction of the other. The near extinction of Old Catalan in Alghero in its competition with modern Italian is a representative example of this situation. One can also see that changing the value of S when a language is in its way to extinction can lead to its recovery. A possible example of this situation is the recovery in recent times of the use of Quechua in its competition with Spanish in Peru.

[3] An applet can be found at: http://ifisc.uib.es/eng/lines/complex/APPLET_LANGDYN.html.

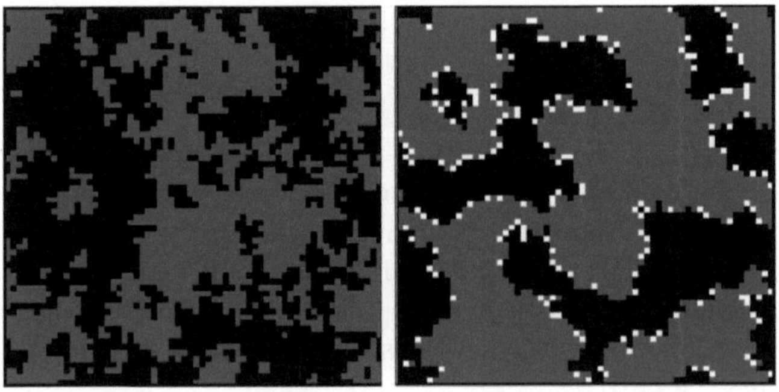

Fig. 3.1 Snapshots showing the formation of domains in the AS-model (*left*) and the MW-model (*right*) starting from an initial random distribution of states of the agents. Neutral volatility ($a = 1$) and socially equivalent languages ($S = 0.5$). $N = 64^2$ agents. Snapshots at time $t = 200$. Notice that in the MW-model, bilingual agents do not form domains, but they place themselves at the interfaces between monolingual domains. *Grey* monolingual X; *black* monolingual Y; *white* bilingual Z

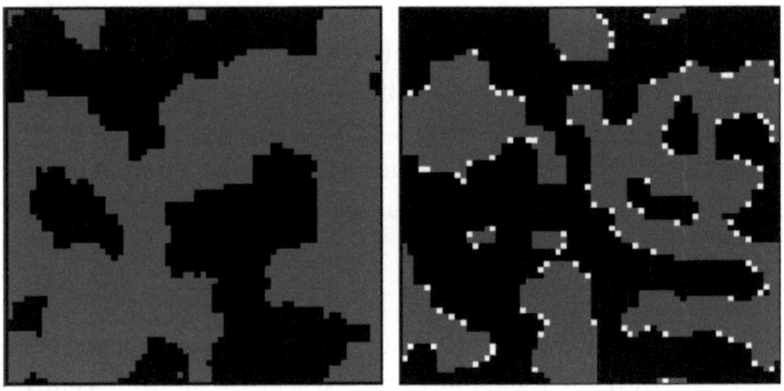

Fig. 3.2 Snapshots showing the formation of domains in the AS-model (*left*) and the MW-model (*right*). Low volatility ($a = 3$) and socially equivalent languages ($S = 0.5$). $N = 64^2$ agents. Snapshots at time $t = 350$. Notice that the boundaries are flatter, due to the increase of curvature driving. *Grey* monolingual X; *black* monolingual Y; *white* bilingual Z

- Low volatility regime ($a > 1$)

  When volatility is low, i.e. agents have larger inertia to change the language they are currently using, both models display a similar growth of monolingual domains (see Fig. 3.2). These domains evolve smoothly and slowly (curvature-driven like), and the times for extinction increase: language death becomes a slower process.

  For socially asymmetric languages, low volatility delays the effect of prestige difference, so that an endangered language can persist for a long time. In comparison to the AS-model, it is interesting to notice that bilingual individuals slow down further the extinction of the less prestigious language (see Fig. 3.3). An example of this situation can be the competition between Galician-Spanish in Galicia (NW of the Iberian Peninsula), where the low volatility of the Galician

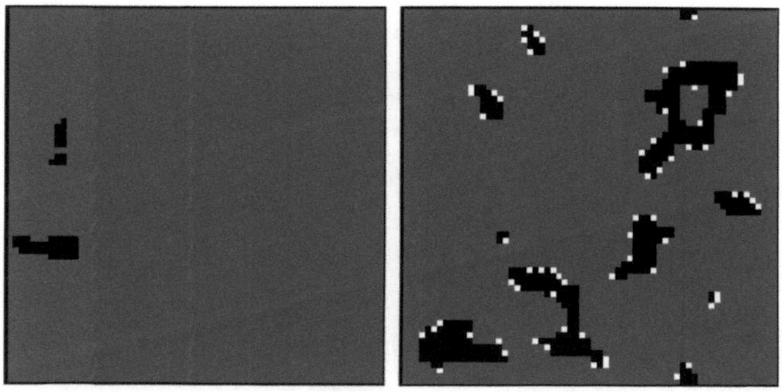

Fig. 3.3 Snapshots in the AS-model (*left*) and the MW-model (*right*). Low volatility ($a = 3$) and socially non-equivalent languages ($S = 0.6$). $N = 64^2$ agents. Snapshots at time $t = 225$. Notice that in the AS-model the less prestigious language is just about to get extinct (around 1% of the population), while in the MW-model the minority language represents still more than 10% of the population. *Grey* monolingual X; *black* monolingual Y; *white* bilingual Z

Fig. 3.4 Snapshots showing the coexistence regime in the AS-model (*left*) and the MW-model (*right*). High volatility ($a = 0.1$) and socially equivalent languages ($S = 0.5$). $N = 64^2$ agents. Snapshots at time $t = 200$. Notice that agents do not form linguistic domains, but are completely mixed. *Grey* monolingual X; *black* monolingual Y; *white* bilingual Z

speakers seems to be preventing a more effective result of current linguistic policies (Monteagudo and Lorenzo 2005), but there are reasons to think that it also prevented Galician from endangerment in the past (Ayestaran Aranaz and Justo de la Cueva 1974).

- High volatility regime ($a < 1$)

  In the case in which volatility is high and for socially equivalent languages ($S = 0.5$), language domains cease to be formed and agents in different states are mixed throughout the population: this scenario leads to a long lived dynamical coexistence of the two languages in both models, with the two languages having the same proportion of speakers and also the survival of a large number of bilingual agents in the MW-model (see Fig. 3.4). The high frequency of changes in the language used by the agents makes possible a linguistic interpretation

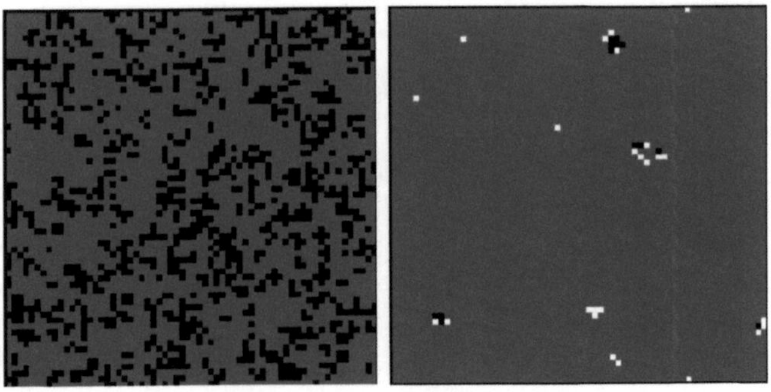

Fig. 3.5 Snapshots in the AS-model (*left*) and the MW-model (*right*). High volatility ($a = 0.1$) and socially non-equivalent languages ($S = 0.6$). $N = 64^2$ agents. Snapshots at time $t = 40$. Notice that in the MW-model the less prestigious language is just about to become extinct, while in the AS-model coexistence is possible. However, this language becomes the one spoken only by a minority. *Grey* monolingual X; *black* monolingual Y; *white* bilingual Z

of this phenomenon as *code-switching*: all agents in the lattice shift languages so often that they can be considered to do so even within one single speech exchange. Examples of this sociolinguistic situation in which agents tend to develop a linguistic variety in which they merge both languages in their speech are the case of *Yanito* spoken in Gibraltar (UK colony on the south of the Iberian Peninsula) and the use of *Spanglish* in certain areas of the USA.

The situation is different when languages with different prestige are considered in a situation of high volatility ($S \neq 0.5$; see Fig. 3.5). For a relatively small difference in prestige between the two languages ($S = 0.6$), bilingual agents in the MW-model cause a fast extinction of the less prestigious language, while in the absence of bilingual agents (AS-model) both languages coexist for long times (although the majority uses the more prestigious language, around 70% of the population). When the prestige difference becomes larger ($S \geq 0.7$), the less prestigious language dies out in both models rather fast (but still slower when there are no bilingual agents (AS-model)).

In summary, numerical simulations of the AS and WM models show that depending on the volatility of individuals and the relative difference on prestige between both languages, the population can either remain indefinitely in a *coexistence* state with a finite fraction of speakers in each of the two languages, or it can reach a *dominance/extinction* state in which one of the two languages takes over the whole population. Our results make clear that prestige is very important, but it is not the whole story, volatility being a very important social parameter in language competition. For example, when a language becomes extinct, this happens much faster in the high volatility regime than in the low volatility regime (compare Figs. 3.3 and 3.5). Generally speaking, high volatility is good for the coexistence of languages of similar prestige. However, when a language is situated in a low prestige position, low volatility of the agents gives larger times before extinction, and in this way, enough time to try to enhance its prestige. This delay in the path to extinction is reinforced by the presence of bilingual agents (MW-model). At the point in which social equivalence is achieved, if the volatility is increased, a situation of coexistence for both languages becomes viable and can be maintained indefinitely.

We finally note that our analysis is here based on an underlying regular two-dimensional network of interactions. This set-up accounts for the important ingredient of local interactions, but other dynamical phenomenology such as the existence of metastable states (coexistence for finite but long times) appears in more complex social networks with community or mesoscale structure (Castelló et al. 2007; Toivonen et al. 2009). In these topologies, the agents form language communities which are correlated with the network structure.

3.3 Macroscopic Descriptions of IBMs of Language Competition

A macroscopic description of the IBMs can be given in terms of the dynamic evolution of the global densities x and y of X and Y speakers, respectively (with the density of bilingual agents being $z = 1 - x - y$), and of the time dependence of the density of pairs of neighbours in a different state ρ. The quantity ρ describes the linguistic boundaries. The case $x = 1$ or $x = 0$ together with $\rho = 0$, corresponds to the *dominance* or *extinction*, respectively, of language X, with all individuals using the same language; while $0 < x < 1$ and $\rho > 0$ indicates that both languages are present in the system (*coexistence*). The aim is to derive, from the IBMs, macroscopic equations for the time evolution of x, y and ρ, and to analyze their predictions for the evolution of the system. These equations depend on the underlying network of interactions, and we consider here the AS-model in different network topologies: we start from the case of a highly connected society with no social structure (fully connected network), that corresponds to the simplified assumption of a "well mixed" population, widely used in population dynamics and language dynamics (Stauffer et al. 2007). To account for the local effects of social interaction among individuals we consider a complex network of interactions. We will see that the results depend on the particular properties of the network under consideration, reflected in the statistical properties of the distribution of the number of links per node of the network. In order to make further contact with the analysis of IBMs of the previous section, and to account for processes of linguistic domain growth, we also consider an approximate description of the dynamics in a regular two-dimensional network by means of a continuous space-dependent field for the density of X speakers. Finally, the effect of bilingual agents, as considered in the MW-model, is discussed in the case of a fully connected network.

3.3.1 *The Abrams–Strogatz Model*

3.3.1.1 Mean Field Description of Fully Connected Networks

We consider a network with N nodes in which each node has a connection (link) to any other node. In a time step $\delta t = 1/N$, a node i with state $X(Y)$ is chosen with probability $x(y)$. Then, according to the transitions (3.1), i switches its state with probability:

$$p_{i,X\to Y} = (1-S)y^a,$$
$$p_{i,Y\to X} = Sx^a, \tag{3.4}$$

where, in this fully connected network, the densities of neighbours of i with states $X(Y)$ are equal to the global densities x and y of nodes in states X and Y, respectively. With these switching probabilities it is easy to obtain that (see Appendix 1):

$$\frac{dx}{dt} = x(1-x)\left[Sx^{a-1} - (1-S)(1-x)^{a-1}\right]. \tag{3.5}$$

Equation (3.5) describes the evolution of the density of X-speakers in a very large population ($N \gg 1$), neglecting finite size fluctuations. The density ρ is not an independent quantity in a fully connected network. It can be obtained as the ratio between total number of links between nodes in different state and the number of links in the network. For large N it becomes

$$\rho(t) = 2x(t)\left[1 - x(t)\right]. \tag{3.6}$$

Equation (3.5) has three stationary solutions

$$x = 0, \quad x^*(a,S) = \frac{(1-S)^{\frac{1}{a-1}}}{(1-S)^{\frac{1}{a-1}} + S^{\frac{1}{a-1}}} \quad \text{and} \quad x = 1.$$

The solutions $x = 1$ and $x = 0$ correspond to the complete dominance of X and Y speakers respectively, while $x^*(a,S)$ is a solution corresponding to a coexistence of X and Y speakers with relative fractions x^* and $1-x^*$, respectively, that depend on a and S. Because we are looking for a long term stationary state, the corresponding solution must be stable under small perturbations (variations in the densities x and y). If the perturbation dies out we say that the solution is stable, otherwise if it grows in time then the solution is unstable. The stability of each stationary solution depends on the values of the parameters a and S, as we summarize in Fig. 3.6. For $a < 1$, the coexistence solution $x^*(a,S)$ is stable. In this coexistence region, and given that large values of S favour X speakers, when $S > 0.5$ then $x^* > 0.5$ and vice-versa. For $a > 1$, the stable solutions are those of dominance

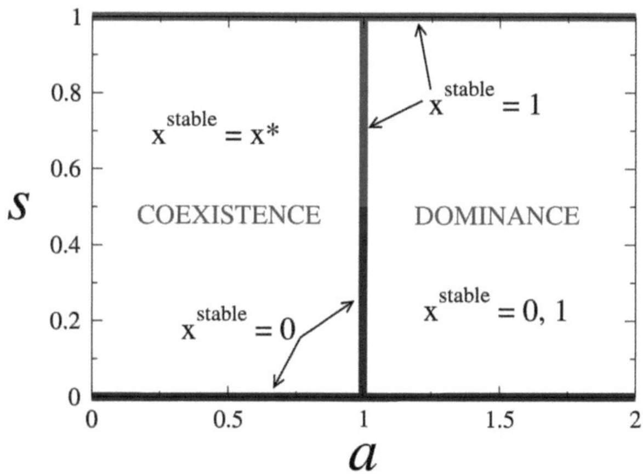

Fig. 3.6 Coexistence and dominance regions of the Abrams–Strogatz model in a fully-connected network. For values of the volatility parameter $a > 1$, the stable solutions are those of language dominance, i.e., all individuals using language X ($x = 1$) or all using language Y ($x = 0$), whereas for $a < 1$ both languages coexist, with a relative fraction of speakers that depends on a and the prestige S. In the extreme case $S = 1$ ($S = 0$), only language switchings towards X (Y) are allowed, and thus only one dominance state is stable, independent of a

($x = 1$) and extinction ($x = 0$). In summary, the fact that agents are highly volatile for $a < 1$, favours language coexistence, and on the contrary, in the low volatility regime $a > 1$, the final state is one of dominance/extinction.

We note that for neutral volatility $a = 1$, the Abrams–Strogatz model becomes equivalent to the biased voter model (Vazquez and Eguiluz 2008; see Appendix 1). In finite systems, the ultimate state is always the dominance of one language. If $S > 1/2$ ($S < 1/2$), language X (Y) dominates, while for $S = 1/2$, the probability that a given language dominates equals the initial fraction of speakers of that language.

3.3.1.2 Complex Networks

In real life, most individuals in a large society interact only with a small number of acquaintances. Therefore, we consider a network of N nodes, with a given degree distribution P_k, representing the fraction of individuals connected to k neighbours, such that $\sum_k P_k = 1$. In order to develop a mathematical approach that is analytically tractable, we assume that the network has no degree correlations, as it happens for instance in Erdös-Renyi and scale-free networks (Albert and Barabasi 2002).[4] Therefore, we can see the system as composed by a collection of nodes

[4]This approximation is called an *homogeneous pair approximation*. Even that our approach can be improved further in scale-free networks by using an *heterogeneous pair approximation*, we have decided to stick to an homogeneous pair approximation because it is possible, in some cases, to

characterized only by its degree k (number of neighbours) and state X or Y, so that nodes with the same degree and state are considered to be indistinguishable. In a time step $\delta t = 1/N$, a node i with degree k and state X (Y) is chosen with probability $P_k\, x$ ($P_k\,(1-x)$), and then, according to transitions (3.1), i switches its state with probability

$$\begin{aligned} P(X \to Y) &= (1-S)\left(n_y/k\right)^a, \\ P(Y \to X) &= S\,(n_x/k)^a, \end{aligned} \tag{3.7}$$

where we denote by $n_y(n_x)$ the number of neighbours of i in the opposite state $Y(X)$ $(0 \le n_x, n_y \le k)$.

Using these switching probabilities one can write down coupled equations for x and ρ. In general, these are complicated equations (see Appendix 1). As an example, the equations obtained for neutral volatility $a = 1$ are:

$$\frac{dx}{dt} = (2S-1)\frac{\rho}{2} \tag{3.8}$$

$$\frac{d\rho}{dt} = \frac{\rho}{\mu}\left\{\mu - 2 + \frac{(\mu-1)\,[S-1+(1-2S)x]\,\rho}{x(1-x)}\right\}. \tag{3.9}$$

Local effects are included in these equations through the parameter μ that measures the average number of neighbours of a node in the network: $\mu = \sum_k kP_k$. Equations (3.8–3.9) have three stationary solutions: the extinction and dominance solutions ($x = 0, 1$) and an extra coexistence solution $x = x^*$. One can verify numerically that these three solutions exist generally for different types of networks. Numerical integration of the general equations for different values of a and S allows us to obtain these stationary solutions and analyze their stability. In Fig. 3.7 we plot the resulting stability diagram on the (a, S) plane for a degree-regular random network, that is, a network in which each node is randomly connected to a fixed number of μ neighbours (continuous lines: $\mu = 3$, dashed lines: $\mu = 10$). The solution $x = 1$ is stable (unstable) for values of S above (below) the curve V_1^{μ} (solid curve), and correspondingly, the solution $x = 0$ is stable (unstable) for S below (above) the curve V_0^{μ} (dashed curve), while x^* is stable in the region where both $x = 0, 1$ are unstable. These results define 4 regions in the parameter space: (1) coexistence, when x^* is stable, (2) dominance, when only $x = 1$ is stable, (3) extinction,[5] when only $x = 0$ is stable, (4) extinction/dominance when both $x = 0$ and $x = 1$ are stable. In regions (2) and (3) the dominant solution is fully

obtain analytical expressions for the time evolution of the different densities, and the results are in quite good agreement with the simulations in the IBMs.

[5] Regions (2) and (3) refer of course to language X. From the point of view of language Y, these regions are naturally reversed.

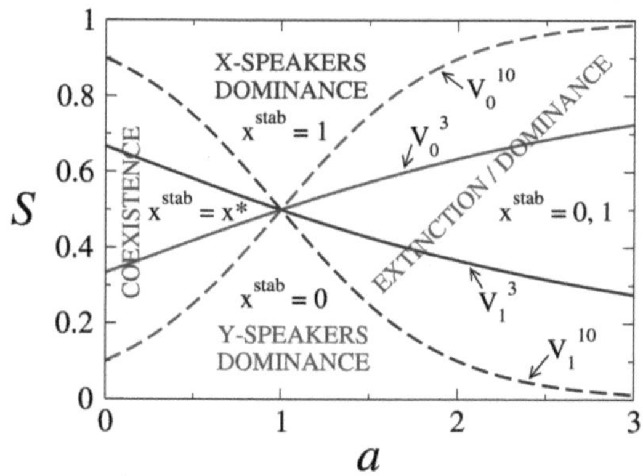

Fig. 3.7 Stability diagram for the Abrams–Strogatz model on a degree-regular random network. *Continuous lines* correspond to $\mu = 3$ and *dashed lines* to $\mu = 10$. In the coexistence region the system is composed of agents using language X or Y, while in the dominance region, users of either one or the other have become extinct, depending on the initial state. We observe that the region of coexistence is reduced, compared to the model on a fully-connected network (Fig. 3.6), and that there are also two single-dominance regions where the same language always dominates

determined by prestige, while in region (4) a solution is chosen also depending on initial conditions.

The case of a fully-connected network, summarized in Fig. 3.6, is recovered in the limit in which the number of links per node approaches the total number of nodes, $\mu \to N$. In this limit, the curves V_0^{μ} and V_1^{μ} approach the step functions in Fig. 3.6. In comparison with that case, we observe that local effects (finite number of neighbours) give rise to the new regions (2) and (3) in which only the most prestigious language is stable. These regions become larger as μ becomes smaller. As a result, region (1) gets shrunk, so that language coexistence is found to be harder to achieve in social networks with low connectivity. This is probably due to the fact that, in networks with low degree and for non-equivalent languages ($S \neq 0$), agents using the more prestigious language reduce their interaction with the minority using the endangered one (in comparison to the case of a fully connected network). This allows for the formation of domains which eventually grow and make it possible for the prestigious language to ultimately dominate the system. In summary, local interactions are predicted to prevent language coexistence and to reinforce the role of prestige.

3.3.1.3 Regular Two Dimensional Network

The behaviour of the Abrams–Strogatz model on a regular two-dimensional network (square lattice), as described qualitatively in Sect. 3.2, is different from its behaviour in fully connected or complex networks. There are two main reasons for these differences. First, the local interactions have an essential role, which is not accounted for in a mean field approximation appropriate for fully connected networks. And

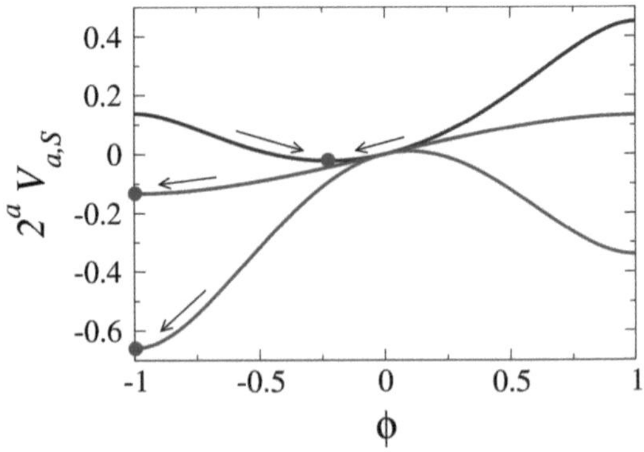

Fig. 3.8 Ginzburg–Landau potential for the Abrams–Strogatz model with prestige $S = 0.6$ and values of volatility $a = 0.1, 1.0$ and 3.0 (from *top* to *bottom*). Potentials are multiplied by the factor 2^a to show them in the same scale. *Arrows* show the direction of the field towards the stationary solution (*solid circles*). For $a = 0.1$ the minimum is around $\phi \simeq -0.25$, indicating that the system relaxes towards a partially ordered stationary state (coexistence), while for $a = 1.0$ and 3.0, it reaches the complete ordered state $\phi = -1$ (dominance)

second, correlations between second, third and higher order nearest-neighbours are important in lattices and were neglected in our previous discussion of complex networks. Such correlations are essential in the formation and growth of the spatial domains discussed in Sect. 3.2. In order to characterize such phenomena, one needs to go beyond the mean field and pair approximations used above. We report here the results of a different approach (Vazquez and López 2008) based on the derivation of a macroscopic equation for a continuous field $\phi_{\mathbf{r}}(t)$ accounting for a space and time coarse grained evolution of the density of users of a language. At the spatial point $\mathbf{r}$, ϕ varies continuously ($-1 < \phi < 1$) so that $\phi = -1$ corresponds to local dominance of language X, $\phi = 0$ corresponds to coexistence with equal strengths of local populations of users of language X and Y and $\phi = +1$ corresponds to local dominance of language Y.

It can be shown (see Appendix 2) that the time evolution of $\phi_{\mathbf{r}}(t)$ can be written in the form of a time dependent Ginzburg–Landau equation

$$\frac{\partial \phi_{\mathbf{r}}(t)}{\partial t} = D(\phi_{\mathbf{r}})\Delta\phi_{\mathbf{r}} - \frac{\partial V_{a,S}(\phi_{\mathbf{r}})}{\partial \phi_{\mathbf{r}}}. \tag{3.10}$$

This can be thought of as a reaction-diffusion equation. The diffusion term, with diffusion coefficient $D(\phi_{\mathbf{r}})$ in front of the Laplacian operator Δ, accounts for local spatial coupling, while the reaction term accounts for global overdamped motion in the potential $V_{a,S}(\phi_{\mathbf{r}})$. The form of this potential gives a basic understanding of the qualitative role of the prestige and volatility parameters previously discussed in the simulations reported in Sect. 3.2.

From the general form of $V_{a,S}(\phi_{\mathbf{r}})$ shown in Fig. 3.8, the dominant effect of prestige becomes clear: For the asymmetric prestige case $S \neq 1/2$ the ordering

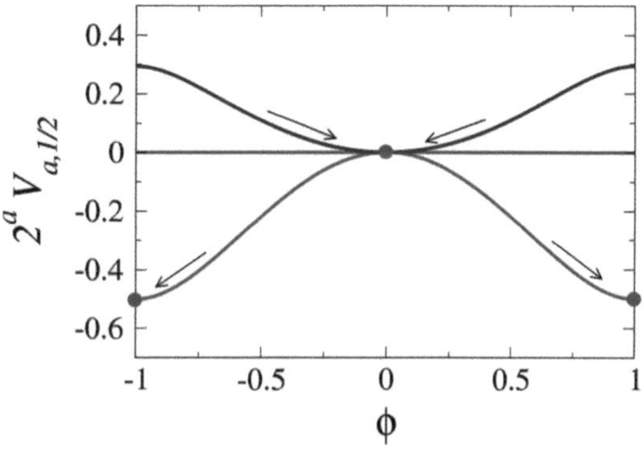

Fig. 3.9 Ginzburg–Landau potential (3.11) for the symmetric case $S = 1/2$ of the Abrams–Strogatz model, with volatility values $a = 0.1, 1.0$ and 3.0 (from *top* to *bottom*). For $a = 0.1$ (high volatility) the system relaxes to a state of coexistence with the same fraction of users of language X and Y uniformly distributed over the space, corresponding to the minimum of the potential at $\phi = 0$, while for $a = 3.0$ (low volatility) it reaches a dominance/extinction state, described by the minima of the field at $|\phi| = 1$

dynamics is strongly determined by S. When $a > 1$ (low volatility), $V_{a,S}$ has the shape of a double-well potential with minima at $\phi = \pm 1$, and with a well deeper than the other. Thus the system is quickly driven by dominant prestige towards the lowest minimum, reaching the dominance state in a rather short time. This is the situation seen in the left panel of Fig. 3.3. On the other hand, for $a < 1$ (high volatility) there is a minimum at $|\phi| < 1$, thus the system relaxes to a state of language coexistence with unequal number of users of language X and Y as the one seen in the left panel of Fig. 3.5.

In the case of symmetric prestige $S = 1/2$ the explicit form of the potential is (see Fig. 3.9)

$$V_{a,1/2}(\phi_{\mathbf{r}}) = 2^{-a}(a-1)\left\{-\frac{\phi_{\mathbf{r}}^2}{2} + [6-(a-2)(a-3)]\frac{\phi_{\mathbf{r}}^4}{24} + (a-2)(a-3)\frac{\phi_{\mathbf{r}}^6}{36}\right\}. \tag{3.11}$$

In this case, with a neutral role of prestige, the role of the volatility parameter becomes even more clear: when $a < 1$ (high volatility) the system relaxes to the minimum of the potential at $\phi = 0$. In this minimum, the average field in a small region around a given point $\mathbf{r}$ is zero, indicating that the system remains in a coexistence state with the same average number of users of language X and Y randomly distributed in space. This is the situation observed in the left panel of Fig. 3.4 after the system has reached a stationary active configuration. For $a > 1$ (low volatility) the potential has two wells with minima at $\phi = \pm 1$ corresponding to the states of dominance or extinction, but with the same depth. Thus there is no preference for any of the two states, and either minimum of the potential is achieved through the formation and growth of linguistic domains: small domains tend to shrink and disappear while large domains tend to grow, reducing the curvature of the linguistic boundaries (Castelló et al. 2006). This situation is the one observed in the simulation in the left panel of Fig. 3.2. For the special case $a = 1$ (neutral volatility) the potential is flat: $V_{1,1/2} = 0$. There is still growth of linguistic domains but now the motion of linguistic boundaries is driven by noise, as observed in the left panel of Fig. 3.1.

3.3.2 The Bilinguals Minett–Wang Model

We consider now the main effect of introducing bilingual agents by studying the MW-model in the simplest case of a fully connected network. In these networks, local densities of neighbours in the different states agree with their global densities. Thus, using the transition probabilities of (3.2) and (3.3), the rate equations for the global densities x and y can be written as

$$\frac{dx}{dt} = Sz(1-y)^a - (1-S)xy^a,$$
$$\frac{dy}{dt} = (1-S)z(1-x)^a - Syx^a, \tag{3.12}$$

where the global density of bilingual agents is $z = 1 - y - x$. The mathematical analysis of these equations is more conveniently done choosing $m = y - x$ and z as independent variables.

One can verify that the points $(m = \pm 1, z = 0)$ in the (m, z) plane are two stationary solutions corresponding to extinction/dominance. But there is also a third non-trivial stationary solution of coexistence, that for the symmetric case $S = 1/2$ occurs for $m = 0$ and a particular value of $z = z^*$. As in the Abrams–Strogatz model, we expect that for a given S, a transition appears at some value a_c of the volatility parameter, where the stability of the stationary solutions changes. By doing a small perturbation around the coexistence solution $(0, z^*)$ in the z direction, one finds that this solution is stable for all values of a. Instead, the stability in the m direction changes at some value a_c which is found to be determined by

$$a_c \ln\left(\frac{1-a_c}{a_c}\right) = \ln\left(\frac{2a_c - 1}{1 - a_c}\right), \tag{3.13}$$

whose solution is $a_c \simeq 0.631$. Then, as in the Abrams–Strogatz model, the (a, S) plane is divided into two regions, but the value of the neutral volatility $a = 1$ is here replaced by a_c. In the high volatility region $a < a_c$, the coexistence solution $(0, z^*)$ is stable, while in the low volatility region $a > a_c$, the stable solutions $(\pm 1, 0)$ indicate the ultimate dominance of one of the languages and the extinction of the other. Since the transition value $a_c \simeq 0.631$ is smaller than the value $a_c = 1$ for the Abrams–Strogatz transition, the region for coexistence is reduced. This has a striking consequence. Suppose that there is population with no bilingual agents, and characterized by a volatility $a = 0.8$, that allows the stable coexistence of the two languages. If now the behaviour of the individuals is changed, so that there exist bilingual agents, language coexistence is lost and finally the system approaches a state with complete dominance of one language. In other words, bilingual agents hinder language coexistence.

3.4 Viability and Resilience of Languages in the Abrams–Strogatz Model

In the previous section, we showed that the mean field description for the AS–model (see (3.5)) predicts (1) the extinction of one of the language for $a \geq 1$, whatever the value of prestige S is, and (2) the safe coexistence of a bilingual society when $a < 1$, except when one language is already extinct. Nevertheless, Abrams–Strogatz's paper finished with the following remarks (Abrams and Strogatz 2003):

> Contrary to the model's stark prediction, bilingual societies do, in fact, exist. [...] The example of Quebec French demonstrates that language decline can be slowed by strategies such as policy-making, education and advertising, in essence increasing the status of an endangered language. An extension to [the model] that incorporates such control on s through active feedback does indeed show stabilization of a bilingual fixed point.

In this section, we give evidence of this statement by introducing the institutional capacity to modify the prestige of one language. We consider three values of the volatility parameter: $a = 0.2$, 1 and 2. Note that in the case $a = 0.2$ (in general for $a < 1$), the fixed point corresponding to coexistence of the two languages is stable, and thus no control parameter on S needs to be included to stabilize a bilingual fixed point. However, when the difference in the prestige of the two languages is very large, the fixed point might lay outside the constraint set, leading to a situation of coexistence with one of the languages close to extinction (a situation that we may want to avoid).

3.4.1 Language Viability

3.4.1.1 Stating the Viability Problem

In general, when defining the viability constraint set in the case of language competition, in order to characterize a language as endangered, the fraction of people speaking it is not enough: other crucial aspects include the point at which children no longer learn the language as their mother tongue, and the increase of the average age of speakers (in an endangered language, eventually only older generations speak the language). However, these factors are out of the scope of the current approach, and we will assume in this work, as a first approximation, that a fraction of speakers below a critical value becomes an endangered situation. Building up from this point, in the Abrams–Strogatz model, we want to determine all the pair values of density of speakers and language prestige which allow for the coexistence of the two languages. The viability constraint set is defined by setting minimal and maximal thresholds on the density of speakers. Below the minimal threshold, $\underline{x}$, or above the maximal threshold, $\overline{x}$, we consider that language X, or Y respectively, is endangered, meaning that the system is not viable. We set $\overline{x} = 1 - \underline{x}$ such that there is no need to consider explicitly language Y: if the density of speakers x of language X is outside the constraint set, so does the density of speakers of language Y, $1 - x$.

As advocated in Abrams and Strogatz (2003), we introduce prestige S as the control variable. The enhancement of the prestige of an endangered language can be triggered by political actions such as the increase of the prestige, wealth and legitimate power of its speakers within the dominant community, the strong presence of the language in the educational system, the possibility that the speakers can write their language down, and the use of electronic technology by its speakers (Crystal 2000). The computation of the viability kernel for the Abrams–Strogatz model will allow to answer questions like: for a given density of speakers, are there action policies performed in favour of the endangered language that will keep the safe coexistence of the two languages? If the answer is yes, which are convenient policies? The main advantage of using viability theory is that it provides general tools and methods to determine the set of initial densities of speakers for which it is possible to control the system so that the coexistence is ensured.

We suppose here that the prestige can take any value $S \in [0, 1]$ but the action on the prestige is not immediate: the time variation of the prestige $\frac{dS}{dt}$ is bounded by a constant denoted c. This bound reflects that changes in prestige take time: to reach a prestige value S_1 starting from an initial prestige $S_0 < S_1$, the stakeholder will have to anticipate at least $\frac{S_1 - S_0}{c\Delta t}$ time steps, where c is the maximum change per unit time Δt. The viability problem consists of defining a function u of time, which maintains the dynamical system:

$$\begin{cases} \dfrac{dx}{dt} = x(1-x)\left[Sx^{a-1} - (1-S)(1-x)^{a-1}\right] \\ \dfrac{dS}{dt} = u \\ u \in [-c, +c] \;\; ; \; c \in [0, 1] \end{cases} \tag{3.14}$$

inside the viability constraint set K:

$$K = [\underline{x}, \overline{x}] \times [0, 1]. \tag{3.15}$$

Notice that, for simplicity, we illustrate the application of the viability theory using the Abrams–Strogatz model on a fully connected network. The first step is to determine the viability kernel *Viab*(K), defined by all couples (x, S) that are solutions of the system, (3.14), for which there exists at least one control function keeping the system indefinitely inside the viability constraint set defined by (3.15).

3.4.1.2 Computation of the Viability Kernel

We assume that the critical threshold of the density of speakers is 20% of the size of the whole population. Therefore, the viability constraint set is $K = [0.2, 0.8] \times [0, 1]$. The theoretical boundaries of the viability kernel can be computed analytically (Chapel et al. 2010). In addition to the theoretical boundaries, we approximate the viability kernel using the algorithm described in Chap. 7, that considers the dynamics in discrete time Δt. Figure 3.10 shows the analytical and approximated viability kernels of the system for $a = 0.2$, 1, and 2. The line corresponding to the fixed points of the dynamics has been obtained using (3.5).

We set the maximal change of prestige per unit time $c = 0.1$, which means that the time variation of the prestige cannot be higher than 10% in a time step. The figure shows how for states with a low X or Y-speakers density, the prestige associated with this language must be strong enough to maintain viability. In situations where the density of one language is high, smaller values of its associated prestige also give rise to viable situations. On the contrary, non-viable states correspond to situations where the density of one language and its associated prestige are low at the same time. In this case, if the actions in favour of this language come too late, its density of speakers will get below the critical threshold 20% while the other will spread through the majority of the population (above 80%). As a increases, the viability kernel shrinks. Indeed, the higher the parameter a, the more rarely agents change their language (low volatility regime). The impact of the change on the prestige is then lower as a increases, which means that when a language is close to the boundary of the viability kernel, even with the maximal government action, the effect on the density of speakers will be too slow to avoid leaving the viability constraint set. On the contrary, as a decreases, agents are likely to change their language (high volatility regime) and to restore coexistence. Note that for $a = 0.2$, the viability kernel is not the whole constraint set: non-viable states reach a stable fixed point located outside K.

3.4.1.3 Determining Heavy Viable Trajectories

The control procedure models an action to enhance the prestige of an endangered language, and we assume that such an action is costly. Therefore, if among different possible action policies to maintain language coexistence, doing nothing keeps the system in a viable situation, we assume that this strategy will be chosen in order to reduce costs. In other words, we suppose that, if several situations with $-c \leq u \leq c$ lead to viable situations, the best choice is $u = 0$. The principle of the control algorithm is described in detail in Chap. 7. Figure 3.11 presents some examples of trajectories for three different values of a, and the time evolution of the control

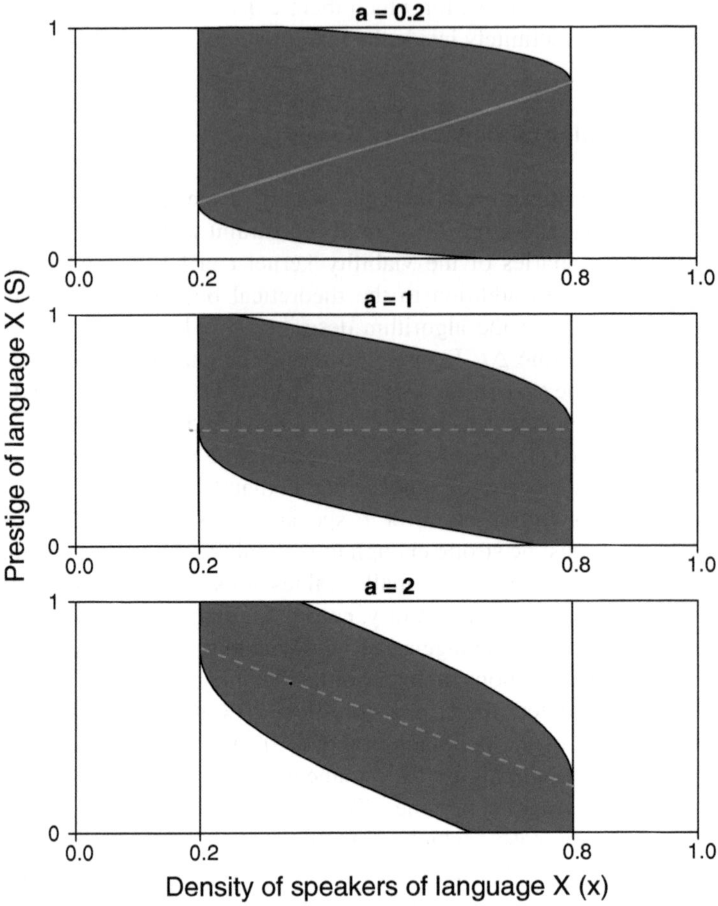

Fig. 3.10 Viability kernel for the Abrams and Strogatz model, with $c = 0.1$ and $\Delta t = 0.05$. From *top* to *bottom*: $a = 0.2$, 1, and 2. The *continuous black lines* represent the theoretical curves of the viability kernel, and the area in *grey* the approximation. The *continuous grey line* represents stable fixed points and the *dotted light grey line* unstable fixed points

($c = 0.1$), during 750 time steps. For $a < 1$, there exist stable fixed points corresponding to coexistence of the two languages and the dynamics settles there, keeping $u = 0$ along the trajectory. For $a \geq 1$, there are no stable fixed points inside the viability kernel, and the control procedure must be applied at each time step. As long as the trajectory is far away from the kernel's boundary, the control is kept to zero; when it approaches the boundary, the control that brings the system away from the boundary corresponds to the maximum value of the control with the appropriate sign, $\pm c$.

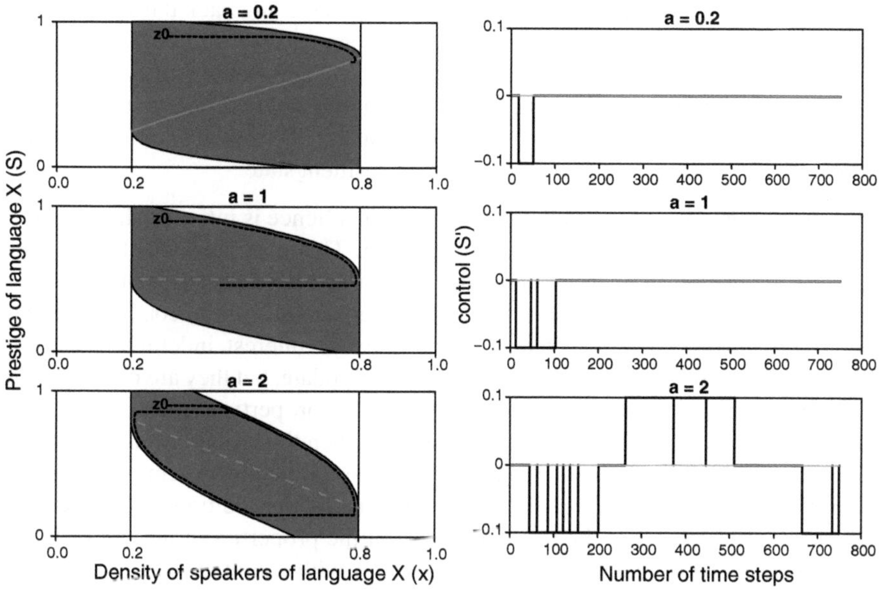

Fig. 3.11 (*Left side*) Examples of trajectories (in *dotted dark grey*) starting from an initial state $z_0 = (x_0, S_0)$ for three values of a ($a = 0.2$, $a = 1$, $a = 2$) and (*right side*) evolution of the control, with $c = 0.1$. The *continuous light grey line* represents stable fixed points and the *dotted light grey line* unstable fixed points

3.4.2 Language Resilience

In the previous subsection, we studied the viability of the AS-model, supposing that one language is endangered when its density of speakers goes below a critical value. However, being endangered does not necessarily mean that the language will disappear. In this section, we are interested in how to maintain or restore coexistence of the two languages when the system is in danger, meaning that a disturbance pulls it outside the viability constraint set.

As we pointed out in the introduction, resilience is the capacity of a system to restore its properties of interest, lost after disturbances. In this section, we define resilience of system (3.14) and (3.15) by considering its capacity to return into its viability kernel when a perturbation pulls it out from it, following Martin's definition of resilience (Martin 2004).

3.4.2.1 Stating the Resilience Problem

We are interested in situations of crisis, which take place when the system leaves the viability constraint set. We distinguish two types of states located outside the viability kernel:

- States for which there exists at least one evolution driving back the system to the viability kernel after leaving the constraint set are called resilient. The system is resilient to a perturbation which leads it into a resilient state
- States for which irrespective of the control policy applied, the system remains outside the viability kernel, are called non-resilient. The system is not resilient to perturbations leading the system into a non-resilient state

For states located inside the viability kernel, the resilience is infinite. Martin (2004) also introduces the notion of cost of restoration. This cost measures the distance between the evolution of the state of the system and the property of interest (i.e. being inside the viability kernel). Its definition must fulfil three conditions. First, the cost of an action which keeps the property of interest indefinitely is zero: maintaining this property may lead to some action update, but they are not taken into account in the cost computation. Second, when the property of interest cannot be restored, the cost of restoration is infinite. Third, when the property can be restored, the cost is finite. It is often defined by the minimum time the system is outside the viability kernel or the minimal deficit accumulated along the trajectory. Then, the resilience is the inverse of the restoration cost of the properties of interest lost after disturbances. The trajectory starting from (x, S) with a minimal cost defines the sequence of "best" action policies to perform, and thus defines the resilience value. Resilience values can be approximated numerically using the algorithm described in Chap. 7. In the context of language competition, the use of viability theory provides a measure of the cost associated to a policy action which will favour an endangered language.

3.4.2.2 Determining the Resilience Basin

All the states can undergo a disturbance. For instance, immigration: people speaking language X are exiled to another country, hence the density of X-speakers reduces dramatically in the home country, and increases in the destination country. Another perturbation to the system can be due to an abrupt change in the prestige of a language because of political actions such as invasion, occupation, etc. The states resulting from disturbances might bring the system outside the constraint set, leading to situations where the density of speakers is lower than the minimal threshold or higher than the maximal threshold. Thus, we consider now the set of all the possible situations $H = [0, 1] \times [0, 1]$, and we study the resilience of the system in H.

First, we determine the set of states of infinite resilience, that are the states located inside the viability kernel of the system defined by (3.14) associated to constraint set defined by (3.15). It corresponds to the dark grey area on Fig. 3.12. Then, we look for all the states for which at least one evolution drives the system back to the viability kernel after spending a finite time in the critical area $H \setminus K$ (where $E \setminus F$ is the complementary set of the set F in the set E). These are the resilient states, represented in light grey in Fig. 3.12. Note that states located in

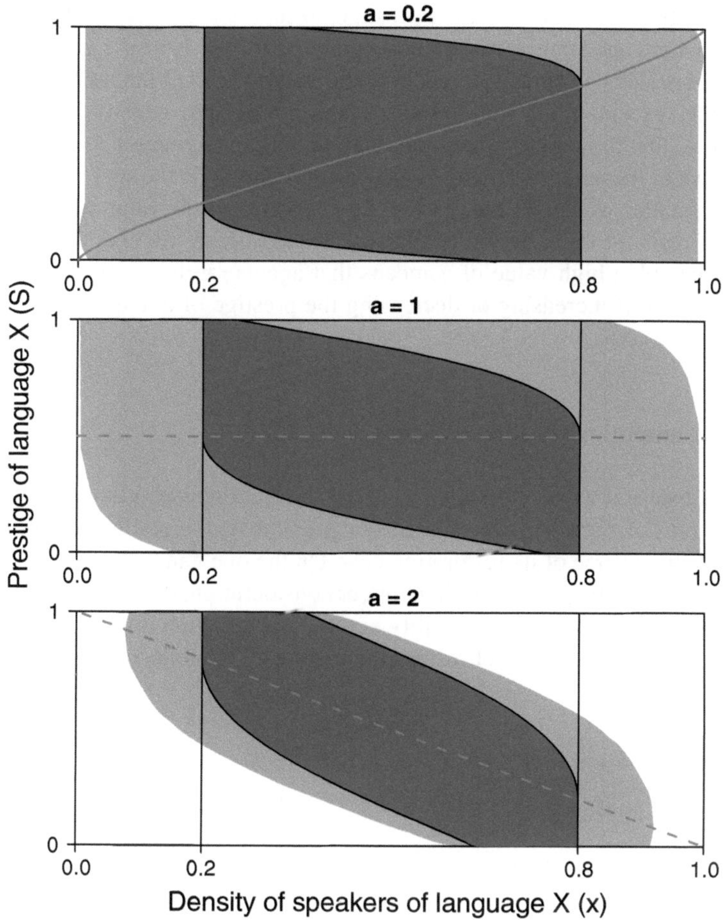

Fig. 3.12 Resilient (in *grey*) and non resilient states (in *white*) in the model associated to the AS-dynamics (3.14) with constraint set (3.15), for three values of a: $a = 0.2$, $a = 1$, $a = 2$. Viability kernel is in *dark grey*

$K\backslash Viab(K)$ can have a finite resilience: when coming back towards $Viab(K)$, the trajectory leaves the constraint set and reaches $Viab(K)$ after spending some time in the critical area. The states that, irrespective of the applied policy, remain outside the viability kernel are in the white zone. For these states, the desired level of language coexistence is impossible and resilience is zero (given the assumed value of c, which limits the effect of action).

In Fig. 3.12, we show the resilient and non-resilient states for $a = 0.2$, 1, and 2. For small values of a, all the states are resilient, except $x = 0$ and $x = 1$, irrespective of the value of S. As we pointed out previously, the fixed point corresponding to coexistence is stable for $a < 1$. Therefore, the desired level of coexistence for the two languages is ensured or can be reached, irrespective of their initial density of speakers and their prestige, except when a perturbation leads to a situation where one language is already extinct. For $a = 1$, nearly for all the initial density of speakers and prestige, reaching the desired level of languages coexistence is possible, except if the initial state represents a large density of speakers of language X that has, at the same time, high prestige (language Y becomes extinct, irrespective of the action applied) or vice versa. For $a > 1$, the set of resilient states becomes smaller, as can be seen in Fig. 3.12. The larger the value of a, the smaller the set of resilient states is. Indeed, as mentioned before for the shrinking of the viability kernel, a high value of a means that agents rarely change their language and the effects of increasing or decreasing the prestige of a language become less effective.

3.4.2.3 Computing Resilience Values

There exist several ways of defining a cost of restoration, depending on the situation and the point of view. As we pointed out previously, the resilience value is then defined as the inverse of its restoration cost. On the one hand, if the time needed to restore viability is the only ingredient under consideration, the cost value is then the time the system is outside the viability kernel. The cost function C_1 that associates to a state x the minimal cost of restoration among all the trajectories starting from z is defined by:

$$\begin{aligned} &C_1(x) = \min_{z(.)}\left(\int_0^{+\infty} \chi_{z(t)\notin Viab(K)} dt\right) \\ &\text{and } \chi_{z(t)\notin Viab(K)} = 1 \text{ when } z(t) \notin Viab(K) \text{ and 0 otherwise,} \end{aligned} \quad (3.16)$$

where z represents the state (x, S), $z(t)$ is the state at time t and $z(.)$ is the trajectory starting from this state. In this way, the cost value is zero when the system is inside the viability kernel. On the other hand, if the cost also depends on how far the system is from reaching the constraint set, the cost function is composed of two terms: the first one that accounts for the time the system is not viable, and the second one, representing the distance to the viability constraint set. In this way, the cost function C_2 associates the time of restoration and the measure of the density of speakers above or below the thresholds of the viability constraint set:

$$\begin{aligned} &C_2(x) = \min_{z(.)}\left(\int_0^{+\infty} \chi_{z(t)\notin Viab(K)} dt + c_2 d(z(t), K)\chi_{z(t)\notin K} dt\right) \\ &\text{and } \chi_{z(t)\notin K} = 1 \text{ when } z(t) \notin K \text{ and 0 otherwise,} \end{aligned} \quad (3.17)$$

where $d(z(t), K) = \max\left(\underline{x} - x(t), x(t) - \overline{x}\right)$ measures the distance between the density $x(t)$ at time t and the density thresholds. Equation (3.17) takes into account that the cost of restoration of a state near extinction is larger than the one for states located near the boundary of K. The parameter c_2 reflects the relative weight of each cost, fixing the cost of being far from K relatively to the time spent outside the viability kernel.

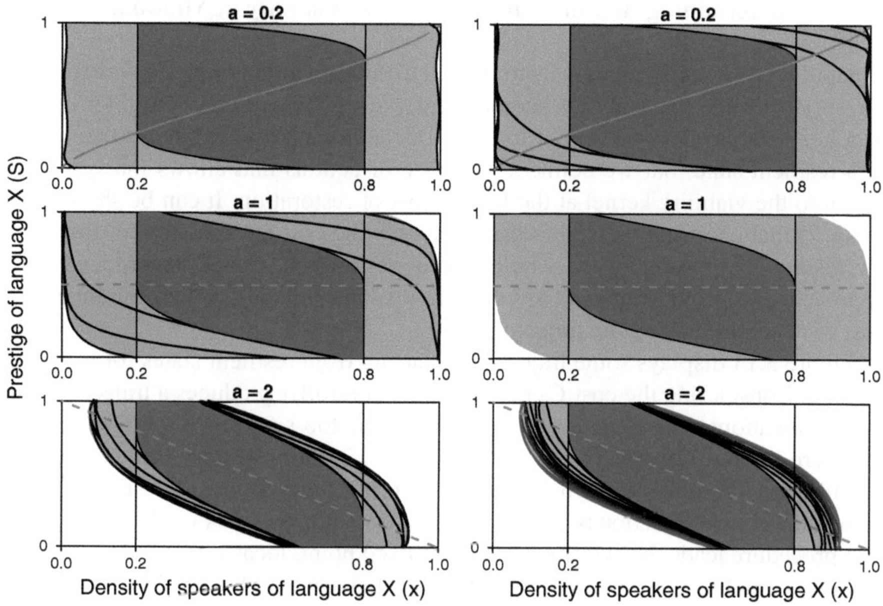

Fig. 3.13 Resilience basins of the Abrams and Strogatz model. In *dark grey*, the viability kernel; between the level lines (*light grey area*), the cost of restoration is finite (one level line corresponds to a cost of 4.8 and the lighter the line, the higher the cost); in the white area, the cost is infinite and the resilience null. (*left side*) cost function defined by (3.16). (*right side*) cost function defined by (3.17)

Figure 3.13 compares resilience basins for the Abrams–Strogatz model for different values of a, and for the two cost functions defined above (with an arbitrary cost parameter $c_2 = 20$ for the second cost function). The difference of cost between two iso-cost curves is 4.8, and therefore the difference in resilience is $\frac{1}{4.8} \approx 0.2$ (the 4.8 value is arbitrary and is linked to the parametrization of the algorithm in Chap. 7). The darker the line, the lower the cost value is (and hence the higher the resilience value). In the white area, cost is infinite, meaning that restoring coexistence of both languages is impossible. For $a = 0.2$, the maximal cost of restoration is equal to 4.8 for cost function C_1 defined by (3.16) and 19.2 for the cost C_2 defined by (3.17). The cost associated to the function defined by (3.17) is bigger than the one associated with (3.16) because it introduces an additional part (the distance to viability) on the final cost. For $a = 1$, the maximal cost of restoration is more important (14.4 for (3.16) and 62.4 for (3.17)). For $a = 2$, the resilient zone is smaller and the costs of restoration are larger (24 for (3.16) and 67.2 for (3.17)). This means that for higher values of a, where the resilient set is smaller, the cost of restoration is larger: there are less resilient situations and the action policies to be performed in order to restore viability are the most costly.

3.4.2.4 Determining Action Policies to Restore Viability at Minimum Cost

Computing resilience values is instrumental to define action policies that drive back the system inside the viability kernel. Here, we use an optimal controller instead of a heavy controller: we do not look for one action policy that keeps the system in a resilient state, but we define a sequence of actions that allows the system to return to the viability kernel at the lowest cost of restoration. It can be shown (see Chap. 7) that choosing the action that decreases the cost at each step (or increases the resilience), minimizes the whole cost of restoration. Hence, theoretically this approach also provides means to compute resilient policies, which minimize the cost of restoration along the trajectory.

Figure 3.14 displays some trajectories starting from resilient states for $a = 0.2$, 1 and 2. Considering the cost C_2 of (3.17), the controller produces a trajectory that avoids situations where the density of speakers is too small or too large, because these are the most costly. Notice that for $a = 0.2$, the trajectory first reaches the equilibrium line outside K, but in order to bring the system inside the viability kernel, the control function is chosen so that it does not get stuck on this fixed point. The procedure leads the system to a second fixed point, located this time inside the viability kernel. Even if the starting point is located inside K but outside the viability kernel (see for example case $a = 1$), the trajectory crosses the viability constraint set before going back to $Viab(K)$, as it is not possible by definition for these states to directly reach the viability kernel.

3.5 Conclusions

In this chapter, we apply the global approach developed in the PATRES project to the particular case of language competition models. We start with individual based models where explicitly represented agents interact and change their practice of language according to the behaviour of their neighbours. With systematic simulations, we explore the richness of dynamical patterns that such models can produce, in particular in the case when the network of interactions is a regular grid. We found that the parameter called 'volatility', expressing the propensity of agents to change the language they are currently using, is particularly sensitive: the dynamical patterns can change dramatically when the volatility goes through a critical value. Then we derive macroscopic descriptions of these dynamics, which capture the main features of these patterns. We consider different networks of interactions between the agents for the simplest model (AS), and we propose specific approaches to make this derivation. This step is important to get a better understanding of the patterns. It is also a necessary step in the global approach for computing viability and resilience. Indeed, as explained in more details in Chap. 7, the tool computing viability and resilience requires that the dynamical system is described with a small set of differential equations involving a limited number of variables. We illustrate such a computation on the simplest model of language

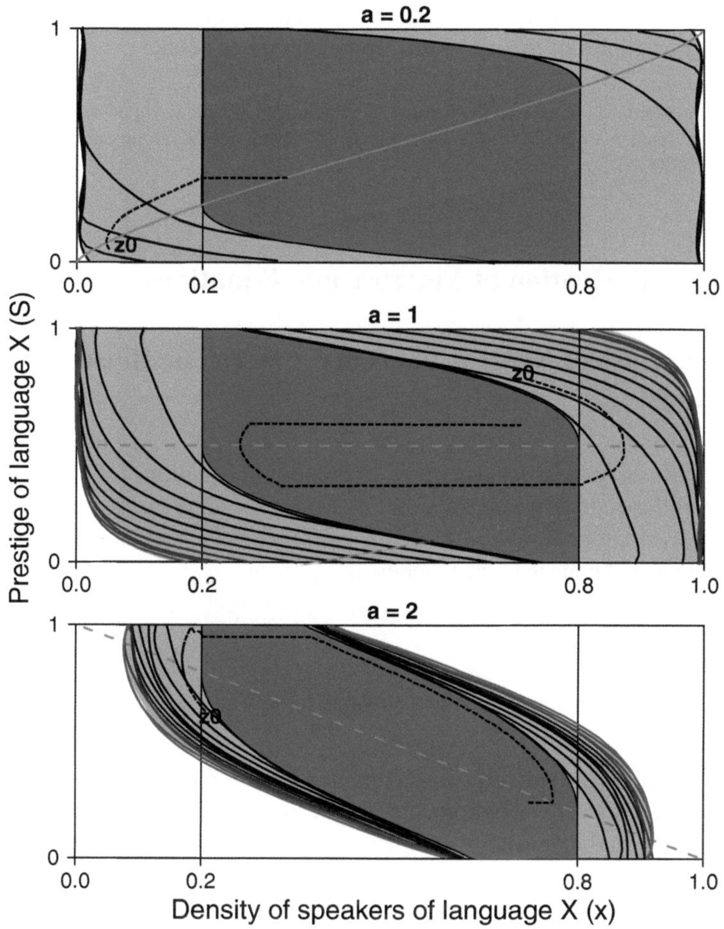

Fig. 3.14 Examples of trajectories (in *dotted dark grey*) starting from a point z_0 during 750 time steps, that allow the system to restore its viability at the minimal cost of restoration, using cost function (3.17). The *continuous grey line* represents stable fixed points and the *dotted light grey line* unstable fixed points. Note that for an initial state $z_0 = (x_0, S_0)$ located inside K but outside $Viab(K)$, the trajectory crosses the viability constraint set boundaries before reaching $Viab(K)$

competition the AS model in the case of total connection. In this setting, we suppose that the prestige of the language can be modified to some extent, as the result of some promotion policy. The analysis shows that the policy to apply for maintaining or restoring the diversity of languages depends heavily on the volatility. When the volatility is bw ($a > 1$), the desired set does not contain any attractor, and we are in a similar case as the last one presented in Chap. 2. It is necessary to act regularly to maintain the balance between the languages, otherwise one language finally gets extinct. In addition, the set of resilient states is smaller. The model of this example is relatively simple, but nevertheless we get a situation where the usual definition of resilience based on attractors would not be applicable. In the next chapters, we study other examples of this approach, applied to other fields.

Acknowledgements L. Loureiro-Porto also acknowledges financial support from the Spanish Ministry of Science and Innovation and European Regional Development Fund (grant no. HUM2007-60706/FILO).

Appendix: Derivation of Macroscopic Equations

A.1 The Case of Neutral Volatility $a = 1$: The Biased Voter Model

In order to illustrate how one can derive and interpret equations for the macroscopic evolution of the system, we consider in this Appendix some mathematical details on different network topologies, with special reference to the simple case $a = 1$ which is also known as the biased voter model.

A.1.1 Fully-Connected Networks

In general, and given switching probabilities $P(X \to Y)$ and $P(Y \to X)$, one has that in the case that the switch occurs, the density x is reduced by $1/N$, thus the average change in the density of X-speakers can be described by the following rate equation

$$\frac{dx}{dt} = \frac{1}{1/N}\left[(1-x)P(Y \to X)\frac{1}{N} - xP(X \to Y)\frac{1}{N}\right]. \tag{3.18}$$

Using the transition probabilities (3.4) in (3.18), one arrives at (3.5). For $a = 1$ and $S = 1/2$, the transition rates in (3.4) become linear in the densities x and y

$$P(X \to Y) = \frac{1}{2}y,$$
$$P(Y \to X) = \frac{1}{2}x. \tag{3.19}$$

Thus apart from the constant prefactor $1/2$, the dynamics is equivalent to adopting the state of a randomly chosen neighbor, that is reminiscent of the voter model (Liggett 1985). If $S \neq 1/2$, the preference for one of the languages makes the voter model to be biased in one direction. For $a = 1$, (3.5) becomes the well known logistic or Verhulst equation

$$\frac{dx}{dt} = (2S - 1)x(1 - x), \tag{3.20}$$

whose solution is

$$x(t) = \frac{x_0}{x_0 + (1 - x_0)e^{-(2S-1)t}}, \tag{3.21}$$

with $x_0 = x(t = 0)$. For a uniform initial condition, $x_0 = 1/2$. Thus

$$x(t) = \frac{1}{2}\Big\{1 + \tanh\left[(S - 1/2)t\right]\Big\}, \tag{3.22}$$

and

$$\rho(t) = \frac{1}{2}\Big\{1 - \tanh^2\left[(S - 1/2)t\right]\Big\}. \tag{3.23}$$

The analytical solutions from (3.22) and (3.23) agree very well with the results from numerical simulations of the model with $S \neq 1/2$, for large enough systems. This is so, because finite-size fluctuations effects are negligible compared to bias effects, even for a small bias (Stauffer et al. 2007).

For $S = 1/2$, the bias is exactly zero, and one obtains that in an infinite large network $dx/dt = 0$, thus x and ρ are conserved. However, in a finite network, fluctuations lead the system to one of the dominance states. To find how the system relaxes to the final state, one needs to calculate the evolution of the second moment $\langle x^2 \rangle$, related to the fluctuations in x, where the symbol $\langle\ \rangle$ represents an average over many realizations. This leads to a decay of the average density of opposite-state links of the form (see Vazquez and Eguiluz 2008)

$$\langle \rho \rangle = \langle 2x(1 - x) \rangle = \langle \rho(0) \rangle\, e^{-2t/N}. \tag{3.24}$$

A.1.2 Complex Networks

Given the switching probabilities of (3.7), if the switch occurs, the density x is reduced by $1/N$, while the density ρ changes by $2(k - 2n_y)/\mu N$. Therefore, in analogy to the case of fully connected networks, but now plugging the transition probabilities from (3.7) into (3.18), we write the average change in the density of X speakers as

$$\begin{aligned}\frac{dx}{dt} = &\sum_k \frac{P_k\,(1 - x)}{1/N} \sum_{n_x=0}^{k} B(n_x, k) S \left(\frac{n_x}{k}\right)^a \frac{1}{N} \\ &- \sum_k \frac{P_k\, x}{1/N} \sum_{n_y=0}^{k} B(n_y, k)(1 - S) \left(\frac{n_y}{k}\right)^a \frac{1}{N},\end{aligned} \tag{3.25}$$

and similarly, the change in the density of opposite-state links as

$$\frac{d\rho}{dt} = \sum_k \frac{P_k\,(1-x)}{1/N} \sum_{n_x=0}^{k} B(n_x,k) S \left(\frac{n_x}{k}\right)^a \frac{2(k-2n_x)}{\mu N}$$
$$+ \sum_k \frac{P_k\, x}{1/N} \sum_{n_y=0}^{k} B(n_y,k)(1-S) \left(\frac{n_y}{k}\right)^a \frac{2(k-2n_y)}{\mu N}, \qquad (3.26)$$

where we denoted by $B(n,k)$, the probability that a node of degree k and state $X(Y)$ has n neighbors in the opposite state $Y(X)$.

Defining the a-th moment of $B(n_x,k)$ as

$$\langle n_x^a \rangle_k \equiv \sum_{n_x=0}^{k} B(n_x,k) n_x^a,$$

and similarly for $B(n_y,k)$, we arrive to the equations

$$\frac{dx}{dt} = \sum_k \frac{P_k}{k^a} \left[S(1-x)\langle n_x^a \rangle_k - (1-S)x\langle n_y^a \rangle_k \right], \qquad (3.27)$$

$$\frac{d\rho(t)}{dt} = \sum_k \frac{P_k}{\mu\, k^a} \Big\{ S(1-x)\left[k\langle n_x^a \rangle_k - 2\langle n_x^{(1+a)} \rangle_k\right] +$$
$$(1-S)x\left[k\langle n_y^a \rangle_k - 2\langle n_y^{(1+a)} \rangle_k\right] \Big\}. \qquad (3.28)$$

In order to develop an intuition about the temporal behaviour of x and ρ from (3.27) and (3.28), we analyze the simplest case $a = 1$. A rather complete analysis of the time evolution and consensus times of this model on uncorrelated networks, for the symmetric case $S = 1/2$, can be found in Vazquez and Eguiluz (2008). Following a similar approach, here we study the general situation in which the prestige S takes any value. To obtain close expressions for x and ρ, we use the fact that in uncorrelated networks dynamical correlations between the states of second nearest neighbors vanish, and also the system is "well mixed", in the sense that the different types of links are uniformly distributed over the network. Therefore, we assume that the probability that a link picked at random is of type xy is equal to the global density of xy links, ρ. Then, $B(n_x,k)$ becomes the binomial distribution with

$$P(x|y) = \rho/2y \qquad (3.29)$$

as the single event probability that a neighbor of a node with state y has state x. $P(x|y)$ is calculated as the ratio between the total number of links $\rho\mu N/2$ from nodes in state y to nodes in state x, and the total number of links $Ny\mu$ coming out from nodes in state y. Taking $a = 1$ in (3.27) and (3.28), and replacing the first and second moments of $B(n_x, k)$ by

$$\begin{aligned}\langle n_x\rangle &= P(x|y)k,\\ \langle n_x^2\rangle &= P(x|y)k + P(x|y)^2 k(k-1),\end{aligned}$$

leads to the two coupled closed (3.8)–(3.9) for x and ρ.

Given that $\rho \geq 0$, (3.8) shows that if $S > 1/2$ ($S < 1/2$), x approaches to 1 (0), and ρ goes to zero, as expected. Even though an exact analytical solution of (3.8) and (3.9) is hard to obtain, we can still find a solution in the long time limit, assuming that ρ decays to zero as

$$\rho(t) = A\, e^{-t/2\tau(S)}, \quad \text{for } t \gg 1, \tag{3.30}$$

where A is a constant given by the initial state and $\tau(S)$ is another constant that depends on S. To calculate the value of τ, that quantifies the rate of decay towards the solutions $x = 1$ or $x = 0$, we first replace the ansatz from (3.30) into (3.8), and solve for x. We obtain

$$x = \begin{cases} 1 + (1-2S)\tau(S)\rho & \text{if } S > 1/2;\\ (1-2S)\tau(S)\rho & \text{if } S < 1/2.\end{cases}$$

Then, to first order in ρ

$$x(1-x) = \begin{cases} (2S-1)\tau(S)\rho & \text{if } S > 1/2;\\ -(2S-1)\tau(S)\rho & \text{if } S < 1/2.\end{cases} \tag{3.31}$$

Replacing the above expressions for $x(1-x)$ into (3.9), we arrive to the following value of τ

$$\tau(S) = \begin{cases} \frac{\mu-2S}{2(2S-1)(\mu-2)} & \text{if } S > 1/2;\\ \frac{2(1-S)-\mu}{2(2S-1)(\mu-2)} & \text{if } S < 1/2.\end{cases} \tag{3.32}$$

Finally, the fraction of X speakers for long times behave as

$$x = \begin{cases} 1 - \frac{(\mu-2S)A}{2(\mu-2)} \exp\left[-\frac{(2S-1)(\mu-2)}{\mu-2S}t\right] & \text{if } S > 1/2;\\ \frac{[\mu-2(1-S)]A}{2(\mu-2)} \exp\left[-\frac{(1-2S)(\mu-2)}{\mu-2(1-S)}t\right] & \text{if } S < 1/2.\end{cases} \tag{3.33}$$

Using the expression for $\tau(S)$ from (3.32) in (3.31) we find that for $S = 1/2$ is $\rho = \frac{2(\mu-2)}{(\mu-1)}x(1-x)$, in agreement with previous results of the voter model on networks (Vazquez and Eguiluz 2008). Equations (3.30) and (3.33) have the same form as (3.23) and (3.22) in the long time limit, for fully connected networks. We can check that we recover that expressions by taking $\mu = N - 1 \gg 1$ in (3.30) and (3.33). This result means that the evolution of x and ρ in the biased voter model on uncorrelated networks is very similar to the mean-field case, with the time rescaled by the constant τ that depends on the topology of the network, expressed by the mean connectivity μ. From the above equations we observe that the system reaches the dominance state $\rho = 0$ in a time of order τ. For the special case $S = 1/2$, τ diverges, thus (3.30) and (3.33) predict that both x and ρ stay constant over time. However, as mentioned in the previous section, the absorbing state is reached by finite-size fluctuations. Taking fluctuations into account, one finds that the approach to the final state is described by the expression (Vazquez and Eguiluz 2008):

$$\langle \rho \rangle = \frac{(\mu - 2)}{2(\mu - 1)} \left(1 - m_0^2\right) e^{-2t/\tau}, \tag{3.34}$$

where m_0 is the initial magnetization and τ is the relaxation time that depends on the system size and the first and second moments of the degree distribution.

2. Equation for the Field $\phi_{\mathbf{r}}$

In order to obtain an equation for the time evolution of the field $\phi_{\mathbf{r}}(t)$ we use a standard method (Vazquez and López 2008): We can interpret X and Y speakers as particles with spins $s = -1$ (down arrow) and $s = 1$ (up arrow) respectively. In other words, we map the language model into a spin-1/2 model, like the Ising model for ferromagnetism. Then, we define by $\phi_{\mathbf{r}}(t)$ the spin field at site $\mathbf{r}$ at time t, which is a continuous representation of the spin at that site ($-1 < \phi < 1$). This is done by placing Ω spin particles at each site of the lattice, each representing a different realization of the dynamics, and replacing $\phi_{\mathbf{r}}(t)$ by the average spin value $\phi_{\mathbf{r}}(t) \to \frac{1}{\Omega}\sum_{j=1}^{\Omega} S_{\mathbf{r}}^j$, where $S_{\mathbf{r}}^j$ is the spin of the j-th particle inside site $\mathbf{r}$. Within this formulation, the dynamics is the following. In a time step of length $\delta t = 1/\Omega$, a site $\mathbf{r}$ and a particle from that site are chosen at random. The probability that the chosen particle has spin $s = \pm 1$ is equal to the fraction of $\pm$ spins in that site $(1 \pm \phi_{\mathbf{r}})/2$. Then the spin flips with probability

$$\begin{aligned} P(- \to +) &= (1 - S)\left(\frac{1 + \psi_{\mathbf{r}}}{2}\right)^a, \\ P(+ \to -) &= S\left(\frac{1 - \psi_{\mathbf{r}}}{2}\right)^a, \end{aligned} \tag{3.35}$$

where $\psi_{\mathbf{r}} \to \frac{1}{4}\sum_{\mathbf{r}'/\mathbf{r}} \phi_{\mathbf{r}'}(t)$ is the average neighbouring field of site $\mathbf{r}$, and the sum is over the 4 first nearest-neighbours sites $\mathbf{r}'$ of site $\mathbf{r}$. If the flip happens, $\phi_{\mathbf{r}}$ changes by $-2s/\Omega$, thus its average change in time is given by the rate equation

$$\frac{\partial \phi_{\mathbf{r}}(t)}{\partial t} = [1 - \phi_{\mathbf{r}}(t)]\, P(- \to +) - [1 + \phi_{\mathbf{r}}(t)]\, P(+ \to -), \tag{3.36}$$

where the first (second) term corresponds to a $- \to +$ $(+ \to -)$ flip event. We have also rescaled the time by $1/\Omega$. To obtain a closed equation for ϕ, we substitute the expression for the transition probabilities (3.35) into (3.36), and write it in the more convenient form

$$\frac{\partial \phi}{\partial t} = \frac{(1-S)}{2^a}(1-\phi)(1+\psi)(1+\psi)^{a-1} - \frac{S}{2^a}(1+\phi)(1-\psi)(1-\psi)^{a-1}, \tag{3.37}$$

where ϕ and ψ are abbreviated forms of $\phi_{\mathbf{r}}$ and $\psi_{\mathbf{r}}$ respectively. We now replace the neighbouring field ψ in the terms $(1+\psi)$ and $(1-\psi)$ of (3.37) by $\psi \equiv \phi + \Delta\phi$, where Δ is defined as the standard Laplacian operator $\Delta\phi_{\mathbf{r}} \equiv \frac{1}{4}\sum_{\mathbf{r}'/\mathbf{r}}(\phi_{\mathbf{r}'} - \phi_{\mathbf{r}}) = \psi_{\mathbf{r}} - \phi_{\mathbf{r}}$, and obtain

$$\begin{aligned}\frac{\partial \phi}{\partial t} &= 2^{-a}(1-\phi^2)\left[(1-S)(1+\psi)^{a-1} - S(1-\psi)^{a-1}\right] \\ &\quad + 2^{-a}\left[(1-S)(1-\phi)(1+\psi)^{a-1} + S(1+\phi)(1-\psi)^{a-1}\right]\Delta\phi. \end{aligned} \tag{3.38}$$

Using a Taylor series expansions around $\psi = 0$ in the right hand side of (3.38),

$$(1 \pm \psi)^{a-1} = 1 + (a-1)\psi + \frac{1}{2}(a-1)(a-2)\psi^2 + \frac{1}{6}(a-1)(a-2)(a-3)\psi^3 \quad \text{and}$$

$$(1 - \psi)^{a-1} = 1 - (a-1)\psi + \frac{1}{2}(a-1)(a-2)\psi^2 - \frac{1}{6}(a-1)(a-2)(a-3)\psi^3$$

we obtain

$$\begin{aligned}\frac{\partial \phi}{\partial t} &= 2^{-a}(1-\phi^2)\Big\{(1-2S) + (a-1)\psi + \frac{(1-2S)}{2}(a-1)(a-2)\psi^2 \\ &\quad + \frac{1}{6}(a-1)(a-2)(a-3)\psi^3\Big\} + 2^{-a}\Big\{[1-(1-2S)\phi]\left[1 + \frac{1}{2}(a-1)(a-2)\psi^2\right] \\ &\quad + (1-2S-\phi)\left[(a-1)\psi + \frac{1}{6}(a-1)(a-2)(a-3)\psi^3\right]\Big\}\,\Delta\phi \end{aligned} \tag{3.39}$$

We finally replace ψ by $\phi + \Delta\phi$ in (3.39) and expand to first order in $\Delta\phi$, assuming that the field ϕ is smooth, so that $\Delta\phi \ll \phi$. Neglecting ϕ^3 and higher order terms in the diffusion coefficient that multiplies the Laplacian, we arrive to an equation for the spin field

$$\frac{\partial \phi_{\mathbf{r}}(t)}{\partial t} = 2^{-a}\left(1-\phi_{\mathbf{r}}^{2}\right)\left\{1-2S+(a-1)\phi_{\mathbf{r}}+\frac{(1-2S)}{2}(a-1)(a-2)\phi_{\mathbf{r}}^{2}\right.$$
$$\left.+\frac{1}{6}(a-1)(a-2)(a-3)\phi_{\mathbf{r}}^{3}\right\}+\frac{a}{2^{a}}\left[1+(1-2S)(a-2)\phi_{\mathbf{r}}\right.$$
$$\left.+\frac{1}{2}(a-1)(a-4)\phi_{\mathbf{r}}^{2}\right]\Delta\phi_{\mathbf{r}}. \tag{3.40}$$

which has the form of (3.10).

References

Abrams DM, Strogatz SH (2003) Modelling the dynamics of language death. Nature 424:900

Albert R, Barabasi AL (2002) Statistical mechanics of complex networks. Rev. Mod. Phys. 74(1):47

Ayestaran Aranaz M, Justo de la Cueva A (1974) Las familias de la provincia de Pontevedra (Galleguidad y conflicto lingüístico). Instituto de Ciencias de la Familia, Sevilla

Bradley D, Bradley M (eds) (2002) Language endangerment and language maintenance. Routledge-Curzon (Taylor & Francis Group), London

Castelló X, Eguíluz VM, San Miguel M (2006) Ordering dynamics with two non-excluding options: bilingualism in language competition. New J. Phys. 8:308

Castelló X, Toivonen R, Eguíluz VM, Saramäki J, Kaski K, San Miguel M (2007) Anomalous lifetime distributions and topological traps in ordering dynamics. Europhys. Lett. 79:66006 (1–6)

Chapel L, Castelló X, Bernard C, Deffuant G, Eguíluz VM, Martin S, San Miguel M (2010) Viability and resilience of languages in competition. Plos one 5(1):e8681

Crystal D (2000) Language death. Cambridge University Press, Cambridge

Fishman JA (1991) Reversing language shift: Theoretical and empirical foundations of assistance to threatened languages. Multilingual Matters, Clevedon

Grenoble LA, Whaley LJ (1998) Toward a typology of language endandergement. In: Endangered languages. Current issues and future prospects. Cambridge University Press, Cambridge

Grenoble LA, Whaley LJ (2006) Saving languages: An introduction to language revitalization. Cambridge University Press, Cambridge/New York

Hinton L, Hale K (2001) The green book of language revitalization in practice. Academic, San Diego

Labov W (1972) Sociolinguistic patterns. University of Pennsylvania Press, Philadelphia

Liggett TM (1985) Interacting particle systems. Springer, New York

Manley MS (2008) Quechua language attitudes and maintenance in Cuzco, Peru. Language Policy 7:323–344

Martin S (2004) The cost of restoration as a way of defining resilience: A viability approach applied to a model of lake eutrophication. Ecol. Soc. 9(2)

Minett JW, Wang WS-Y (2008) Modelling endangered languages: The effects of bilingualism and social structure. Lingua 118:19–45

Monteagudo E, Lorenzo A (directors) (2005) A sociedade galega e o idioma. A evolución sociolingüística de Galicia (1992–2003). Santiago de Compostela: Consello da Cultura Galega

Nettle D, Romaine S (2000) Vanishing voices. The extinction of the world's languages. Oxford University Press, Oxford

Stauffer D, Castelló, X, Eguíluz VM, San Miguel M (2007) Microscopic Abrams-Strogatz model of language competition. Physica A 374:835–842

Toivonen R, Castelló X, Eguíluz VM, Saramäki J, Kaski K, San Miguel M (2009) Broad lifetime distributions for ordering dynamics in complex networks. Phys. Rev. E 79:016109
Tsunoda T (2005) Language endangerment and language revitalisation. Mouton de Gruyter, Berlin and New York
UNESCO (2003) Language vitality and language endangerment. UNESCO (Ad Hoc Expert Group on Endangered Languages)
Vazquez F, Eguíluz VM (2008) Analytical solution of the voter model on uncorrelated networks. New J. Phys. 10:063011
Vazquez F, López C (2008) Systems with two symmetric absorbing states: relating the microscopic dynamics with the macroscopic behavior. Phys. Rev. E 78:061127
Vazquez F, Castelló X, San Miguel M (2010) Agent based models of language competition: Macroscopic descriptions and order-disorder transitions. J. Stat. Mech. P04007
Wang WS-Y, Minett JW (2005) The invasion of language: emergence, change and death. Trends Ecol. Evol. 20:263–269
Wölck W (2004) Universals of language maintenance, shift and change. Coll Antropol 28(1):5–12

Chapter 4
Viable Web Communities: Two Case Studies

Dario Taraborelli and Camille Roth

4.1 The Viability of Online Social Systems

Addressing the question of what makes an online social system "viable" requires some preliminary conceptual clarifications in order to define the scope of the present analysis. Section 4.1 of the present chapter is devoted to framing the problem conceptually: we first introduce the notion of a collaborative Web community by considering the properties that characterise it; we then discuss a number of ways in which the viability of these systems can be defined and the challenges faced by empirical research in identifying measurable indicators of viability. In Sect. 4.2 we present an empirical analysis of two paradigmatic cases of collaborative Web communities and discuss methodological issues emerging from the study of their dynamics from the point of view of viability. We conclude by presenting in Sect. 4.3 a simple model of viable online communities based on the empirical and conceptual considerations of the first two sections. Viability, we argue, is a notion that is hard to frame in the case of social systems. By discussing alternative characterisations of this notion and illustrating how to tackle it empirically, this chapter aims at offering methodological insights into the study of viable online social systems. These insights, we submit, are a precondition for the application of formal definitions of viability and resilience (see Chap. 2 of this volume) to realistic models of how collaborative systems function.

D. Taraborelli
Centre for Research in Social Simulation, University of Surrey, Guildford GU2 7XH, Surrey, United Kingdom
e-mail: d.taraborelli@surrey.ac.uk

C. Roth (✉)
Centre d'analyse et de mathématique sociales, CNRS-EHESS, 54, boulevard Raspail, 75006 Paris, France
e-mail: roth@ehess.fr

G. Deffuant and N. Gilbert (eds.), *Viability and Resilience of Complex Systems*, Understanding Complex Systems, DOI 10.1007/978-3-642-20423-4_4,

4.1.1 *Definitions*

4.1.1.1 Online Societies and Online Communities

In this chapter we will focus on a particular kind of online social systems that we can characterise as content-based, collaborative Web communities.[1] Such communities are defined by the fact that users joining them participate in the collaborative production of content, whether in the form of peer production (such as in the case of *wikis* or *open source development*), collaborative annotation (as in the case of *social bookmarking* and *collaborative filtering*) or media sharing (as in the case of *social media* platforms).

Web-based platforms offer particularly appealing conditions to study the nature and dynamics of collaborative groups for two main reasons. On the one hand, such systems offer the possibility of empirically studying different aspects of user behaviour at a large scale though the extraction of online datasets via dedicated tools, such as programmable interfaces (or APIs).

On the other hand, these systems offer a particularly suitable ground for the purpose of the present discussion as they are often embedded in what we may call an "online user society." Agents in an online society can be characterised as users with a unique online identity. The active user base of platforms such as *Facebook*, *Flickr*, or the *Wikipedia* can be taken as an example of an online society.[2] Users of such platforms can freely participate in discussion and content production, establish links to other members and create and maintain affiliations to the variety of communities that these platforms support.

Systems supporting an online society are ideally designed to allow the researcher to compare and assess the respective performance of communities that, although different in structure and organisation, tap into the same user base. At the microscopic level, one can observe how the social and affiliation network as well as the participative behaviour of individual users evolve over time. At a mesoscopic level, one can observe how communities evolve over time, how they recruit members and how their structure affects participation. At a macroscopic level, one can study the evolution of an online society as a whole. In this sense, communities targeting the same user base can be seen as competing with each other to recruit members, a condition that makes it possible to study aspects of group dynamics that are often inaccessible in offline communities.

---

[1] We will use in what follows "online groups" and "Web communities" as synonyms to refer to social systems that require an explicit act of affiliation for users to become members, as opposed to communities that can be "detected" on the basis of topological properties of a network (see for example Edling 2002; Newman 2006).

[2] The question of the mapping between real, offline identities and online identities in Web communities is beyond the scope of the present discussion. In our work we refer to a member of an online community as a user that can be identified by a unique online identifier, no matter who actually owns and controls that identifier.

4.1.1.2 A Taxonomy of User-Centred Relations

Before discussing the functioning of online communities, we define several distinct classes of relations involving users of these systems. First, members can entertain multiple group affiliations at the same time – given the ease of joining a group, affiliation should not be taken as a straightforward indicator of active participation. Second, members can create explicit relations to other members in the form of "contact" or "friend" links. Third, members contribute *content* to a shared pool of resources maintained by the community; contributed content can consist of text, code patches, media or even, at a more meta level, of annotations which provide information about the other types of items. These classes of user-centred relations involving content, users and groups, indicate the large range of ways in which a collaborative community can evolve and the multiple forms that member participation can take. In order to tackle the question of the viability of online social systems, it becomes particularly important to be able to compare the performance of different communities on the basis of their structure and internal functioning.

4.1.2 Defining and Securing Viability

Collaborative communities built on top of an online user society face a number of risks that potentially threaten their survival. Peer production systems, for one, typically die of inactivity or an insufficient number of valuable contributions or, conversely, whenever quality assessment becomes unmanageable due to content explosion or ineffective measures against vandalism. The governance of such communities has been based so far on best practices and recommendations, as empirical evidence on the impact of specific policies on how these communities evolve in time is still relatively scarce.[3]

Addressing the problem of the viability of these systems and their governance, therefore requires understanding: (1) what characterises these communities as "viable" and (2) once a viable state is explicitly defined – what policies can be devised in order to achieve or maintain this state.

4.1.2.1 Dimensions of Viability

Viability as Membership Growth

The first typical way to assess the performance of a collaborative community consists in looking at the *growth* of the number of its members over time.

[3]For an example of qualitative, community driven attempts at understanding patterns that affect in a positive or negative way the evolution of a collaborative community, see Mader (2007).

Unsurprisingly, growth is one of the main aspects on which the quantitative literature on collaborative online communities has focussed so far (see Almeida et al. 2007; Voss 2005; Godfrey and Tu 2000, 2001; Lam and Riedl 2009; Mislove et al. 2007; Roth et al. 2008). Studying the viability of a collaborative community by taking into account its population growth (and the speed thereof) is a valuable approach as long as growth per se is a desirable feature for the sustainability of the system. In some cases, however, an uncontrolled growth of participants is likely to lead to the breakdown of collaboration. This is particularly sensitive when growth in population and the growth in the content that a community produces start to diverge: we will address this issue later by referring to the notion of the attentional span of the members of a community. Even if we take the absolute growth in members of a community as an indicator of its performance in securing a solid user base, we need to consider how its growth compares to that of other communities, how it relates to the turnover of its members and how it is affected by different processes through which new members are recruited.

- In cases in which membership does not imply exclusive affiliation, the growth of a given community should not (at least not in principle) be directly affected by the **comparative growth rate** of other communities built on the same user base. However, this is unlikely to be the case as, even in an online society with a constantly growing number of users, communities are de facto competing for members. What is crucial is then to understand the nature and effectiveness of processes by which communities manage to secure their membership.
- Two communities may perform equally well in growing their overall membership over time. However, they may differ significantly in how good they are at maintaining existing members, or controlling **member turnover**. The same net growth in population can be the result of (*a*) a slow, steady growth in members without a significant drop-off of existing members or (*b*) the result of a high turnover, whereby the number of new recruits outweighs the number of losses. Considering the role of turnover beside sheer population growth is reasonable if we assume that, while turnover is essential to secure renewal within a community, a high rate of turnover will disrupt the continuity within the community, i.e. a condition that might be needed in order to preserve the possibility of norm transmission within the community (see Forte et al. 2009).
- A final important aspect related to membership growth is the **variety of recruitment processes** that may result in the same observable increase in the population of a given community. The way in which communities recruit members – whether by tapping into the individual social networks of their current members (*social recruitment*) or by focussing on the interests of the new recruits (e.g. *recruitment by homophily*) – has crucial consequences. As we will see in the second part of this chapter, social recruitment processes are likely to reach a point after which the community becomes too cohesive to allow a significant turnover in its membership.

In sum, different aspects in the growth of a community's membership can be relevant to its viability: the rate of population growth over time, the relation between growth in content and growth in population, the rate of member turnover, the mechanisms underlying recruitment and population growth.

Viability as Participation

The second challenge in defining the viability of collaborative online communities consists in understanding what diversifies community membership in terms of individual participation and how this, in turn, affects how a community thrives at a macroscopic level. A community may be said to be viable if it manages to secure a minimum number of participants committed to perform the specific tasks within the community that are essential to the achievement of its goals. As self-allocation of effort is a typical feature of peer production systems (see Benkler 2002), individual incentives to participation play a central role in specifying the conditions that make a community viable. A lack of balance between regular members and administrators, a redundancy of effort by community members, and a lack of participants devoted to quality control and norm enforcement are examples of ways in which patterns of participation can be disruptive for a community striving to achieve a specific goal. A detailed analysis of the drivers of participation in peer production is beyond the scope of the present work, but we should mention three key aspects that are relevant to the problem of the viability of these systems:

- The relation between individual **motivation and participation**. What actually drives users to participate in online peer production is an issue that has been addressed by several authors (Hars and Ou 2002; Benkler 2002, 2006). Status recognition and expertise within a community are key factors in strengthening membership, but to what extent these internal factors are effective in terms of recruitment and task allocation is an issue that has still to be explored. A community is viable if it manages to channel individual motivation in a way that is functional to achieving a proper division of labour.
- The distinction between **passive membership and participation**. Several communities thrive despite a relative low number of active members. This suggests that communities may have a high potential of recruitment even though only a small proportion of their membership is responsible for content production. This is a known property of peer production systems such as wiki-based communities where a majority of participants contributes only marginally to content production as opposed to quality control (Kittur et al. 2007; Roth 2007) and in all those systems where the existence of a community of "lurkers" is a vital condition for the performance of a community (Nonnecke et al. 2006). In this respect, the performance of a mature peer production system may depend on more subtle factors than the sheer proportion of actively contributing members: fighting vandalism, for one, seems to depend more on the number of passive watchers regularly monitoring content than on the small proportion of active contributors.

- The relation between **competitiveness and participation**. One final issue that is key to defining the viability of collaborative systems is understanding to what extent these systems are in mutual competition. Depending on the underlying design, two communities competing on the same topical or social "niche" (McPherson 1983) may thrive independently without affecting each other's performance; conversely, one may observe a migration of users from one community to the other (which will affect the viability of each community if measured by membership growth), or a change in participation rate not involving an actual migration or termination of membership (which will only affect the viability of a community as measured through participation metrics).

Understanding drivers of participation and types of participation is crucial to characterise those communities that may be effective at recruiting members but unable to secure a proper division of labour, such as content production versus content maintenance by active users. In this respect, both role diversification and proportion of active participants represent critical variables to measure the maturity of a community and its potential viability.

4.1.2.2 Achieving Viability

Once we have identified a specific standard of viability as a function of what aspects of an online community we wish to focus on, the next question is what policies are available to effectively achieve and maintain that standard. The question bears on the delicate issue of control policies in peer production systems, or how to devise an appropriate governance model for systems where individual effort is typically self-allocated and in which traditional organisational structure is not applicable (see Benkler 2002; Forte et al. 2009). We review in this section three classes of factors that are instrumental in *controlling* the dynamics of a community towards reaching a viable state.

Viability and Quality Control

The content dynamics of a collaborative community deserve as much attention as membership dynamics from the perspective of its viability. As we suggested, a divergence in the respective growth of content and population may easily lead to a breakdown of collaboration. This makes quality control policies one of the main factors behind the successful performance of collaborative communities, a problem that is particularly sensitive in communities where content is collectively curated in order to meet some shared quality standards (Forte et al. 2009). The most extensively studied case of collaborative quality control in online peer production is *Wikipedia* (see Halim et al. 2009; Stvilia et al. 2008; Suh et al. 2008; Wöhner and Peters 2009; Kittur et al. 2007, 2009) – a case in which the effectiveness of quality control policies, the social processes involved in their enforcement and the general distribution of labour among contributors have been empirically analysed. Factors that may drive a community to achieve a viable state from this perspective include the balance between inclusiveness and quality control: too strict quality control policies may drive away potentially valuable contributors, but the same effect can actually result from a demographic explosion or by loose or poorly effective quality control mechanisms.

Viability and Governance

Despite the fact that collaborative communities and peer production systems are often referred to as systems that accept unconstrained contributions from their members, they often implement forms of "soft governance" and hierarchical organisation that help maintain an active community and focussed content production. Governance-related factors include solutions controlling *content production* (i.e. what kind of content is allowed within the community's product) as well as systems controlling *member affiliation* (i.e. who can join or who can perform specific tasks). In both cases governance solutions can be enforced a priori, by imposing limits onto the production of participation or content production, and/or a posteriori, by removing inappropriate content or removing existing members not complying with community norms.

Viability and Sociability

A third class of factors relates to social interactions, especially in collaborative communities built on top of online networking services. Even though governance measures can exert indirect control on the shape of social interactions, a number of properties of the social network of the members of a group have proved to show an important role in controlling the dynamics of collaborative communities – as a driver of recruitment of new affiliates, or, on the contrary, as an obstacle towards further growth, whenever high social cohesiveness hinders the affiliation of users not belonging to the social neighbourhood of a group. In terms of peer production, we may also expect that groups whose members maintain at the same time too large a number of social ties will start showing symptoms of breakdown in collaboration or in the ability to effectively monitor content production, which in turn will threaten the viability of the system.

4.2 Two Case Studies

In this section we illustrate how the above conceptual distinctions can be put to work in the empirical study of the evolution of collaborative systems. We focus in particular on the relation between growth-related viability and control factors by looking at properties that spur or regulate growth in two paradigmatic cases of collaborative systems: peer production systems and communities in social media.

4.2.1 Peer Production Systems

Wikis are, in a broad sense, websites whose content can be contributed and modified by any user in a collective and collaborative fashion. As such, they represent one of the most prominent examples of Web-based peer production systems. The most famous and possibly the most successful of these websites, *Wikipedia*, has attracted a substantial interest in the research community in recent years (Lih 2004; Anthony et al. 2005; Bryant et al. 2005). The Web has seen, however, several thousands of other wikis thrive and proliferate, with varying degrees of success: some recruit many users, achieving sustainability with established role distributions, frequent updates and efficient measures against vandalism, while others fight to attract contributors. Wiki-based communities can have distinct policies or scope but be equally sustainable, or have identical policies but die for a variety of reasons; all endeavoring to survive within what may be called the "wikisphere".

This first case study consists of an exploratory investigation of some factors likely to account for diverse wiki destiny and viability, in terms of technical, social and structural features. In this context, we understand "viability" as dynamic sustainability of both population and quality content: in other words, a viable wiki should be able to grow in terms of articles and users in such a way that the whole content can be maintained by a sufficient number of users. Our aim is however not to provide a formal definition of an appropriate notion of viability for wiki-based communities, but rather to present a detailed descriptive analysis of the demographic and structural dynamics of a large sample of wikis as an empirical basis for further research. In particular, we discuss these results in light of the role played by governance measure in affecting the viability of these communities, moving beyond the *Wikipedia* case.[4]

4.2.1.1 Wiki Dynamics

Various governance systems and software parameters, i.e. technical and social constraints, define a landscape wherein each online community is settled, grows and lives. How can the growth and evolution of such communities be assessed? As content-based online communities, wikis mainly evolve in two dimensions: (*a*) contributors, who may or may not constitute an active community; as discussed e.g. by Bryant et al. (2005); and (*b*) pages, which may or may not amount to authoritative or useful content; as demonstrated for example by Giles (2005).

[4]For an extensive discussion of results presented in this section, see Roth et al. (2008).

Users and pages are likely to obey a dual dynamic: while more users may contribute to more pages, content proliferation seems to require more attention from users. As a first approximation, it may thus seem judicious to assess the healthiness of a wiki through these variables, taken as demographic indicators for its actual growth and activity.[5]

To our knowledge, the present case study represents an original longitudinal analysis of the content and population dynamics of a large set of wikis. As well as almost always focusing on *Wikipedia*, previous quantitative wiki research has mainly examined the topological structure of underlying interaction or hyperlink networks (Capocci et al. 2006; Zlatic et al. 2006) or article-level features (Brandes and Lerner 2008; Wilkinson and Huberman 2007), with little interest in the specific dynamics of the demographic determinants themselves (with the exception of Kittur et al. 2007 who investigates *Wikipedia*'s demographics of casual *vs*. committed contributors).

Dataset

We constructed a dataset made of simple statistics gathered for a large number of MediaWiki-based wikis,[6] which enabled us to consider the same set of variables across all wikis and make sure these variables were generally available. The data was collected over the period August 2007–April 2008 from a publicly-available database.[7] totalling more than 11,500 wikis. We applied further restrictions on this dataset, as described in Roth et al. (2008).

To sum up, the final, "clean" dataset that we considered for this study is made up of about 360 wikis, all of which have an initial population between 400 and 20,000 users, are *not* hosted at some specific 'wiki farms' that do not report useful data, and which do not have major discontinuities in the daily change of their population or content. As such, we assume this subset to be representative of a homogeneous sample of wikis having a relatively sizeable yet not exceptionally large base of registered users – the latter being a hallmark of typical outliers (such as the English Wikipedia) in the wikisphere. Besides, the exclusion of discontinuously-growing wikis ensures that the observed dynamics are due to genuinely bottom-up user-driven behaviour rather than top-down administrative intervention or external attacks. In this sense, and from a viability theory perspective, we are thereby focusing on the *autonomous dynamics* of such systems.

---

[5]As per our considerations in the first section of this chapter, it should be noted that sheer content growth per se may not be a good indicator of a sustainable wiki, as studies on wiki proliferation also seem to suggest Happel and Treitz (2008).

[6]This initial dataset includes among others a large set of *Wikipedias*.

[7]http://s23.org/wikistats/largest_html.php The database is maintained by a user called "Mutante" who graciously granted us the permission to harvest this data.

Variables

We considered a set of four raw quantitative variables: *population size* (U), measured by the number of *registered* users; *content size* (P), measured by the number of so-called "good" pages (i.e. actual *content* pages excluding default pages created by the wiki engine), hereafter indifferently called "pages", "good pages" or "articles" ; *administrator population* (A), the number of users who are granted "administrator status", i.e. special rights to modify sensitive content and perform maintenance activity; and *editing activity* (E), measured by the total number of edits. We also included one qualitative variable indicating the presence of an access control mechanism: *editing permission* (R), i.e. the possibility of creating a page for unregistered/anonymous users. R is either 1, "anonymous editing allowed", or 0, "registered users only". However simplistic these variables may be, they provide key indicators of the global dynamics of a wiki, and shed light on diverse aspects of its structure and evolution. We collected the values of these variables for each wiki every day and over a period of 250 days, i.e. approximatively 8 months.

Structural Metrics

Wiki dynamics were studied as a function of a number of structural metrics based on the above variables, and that we can broadly categorise in two broad types of independent variables. On the one hand, **descriptive indicators**, i.e. variables on which wiki administrators have no direct control: *user activity* (i.e. the proportion of edits per user E/U), *user density* (i.e. the proportion of users per page U/P), and *edit density* (i.e. the proportion of edits per page E/P). On the other hand, **governance factors**, variables that wiki administrators can directly control: *administrator ratio* (i.e. the proportion of users who are granted administrator status A/U), *administrator density* (i.e. the proportion of administrators per page A/P), *editing permission* (R). See Table 4.1 for a summary.

We subsequently assessed wiki dynamics by comparing their diverse growth paths with respect to a set of independent variables. 'Growth' is defined in terms of population and content size variation (see Table 4.2): user growth G_U (resp. page growth G_P) is the ratio between final and initial populations (resp. content sizes): $G_U = U_{\text{last}}/U_{\text{first}}$ (resp. $G_P = P_{\text{last}}/P_{\text{first}}$). For each continuous variable, instead of carrying out a delicate analysis by dealing with clouds of points, we adopted a more

Table 4.1 Wiki metrics used as independent variables

	U/P	User densitiy
	E/U	User activity
Descriptive metrics	E/P	Edit density
	A/U	Administrator ratio
	A/P	Administrator density
Governance metrics	R	Editing permission

Table 4.2 Wiki growth indicators

	G_U	Population growth (U_{last}/U_{first})
Wiki growth	G_P	Content growth (P_{last}/P_{first})

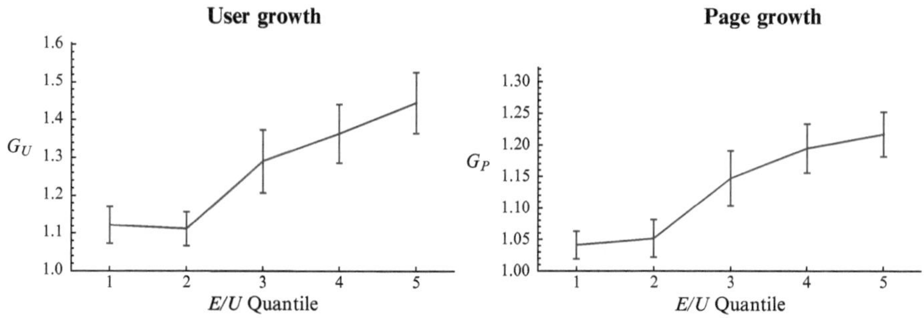

Fig. 4.1 Growth landscape with respect to *user activity*, i.e. the proportion of *edits per user* (E/U).

insightful approach by dividing wikis into five quantiles, each including exactly 20% of all wikis in the clean dataset. We then computed and compared growth rate means over all wikis for each quantile. This representation was applied to all the above-mentioned variables, except for R where there are only two "quantiles" (0 or 1), enabling us to distinguish population quantiles on a unique graph.

4.2.1.2 Determinants of Wiki Dynamics

The results suggest that different structural and governance-related factors have significant correlation with – and plausibly, in some cases, effect on – the content and population dynamics of a wiki:

- **Significant descriptive indicators.** Figure 4.1 shows the effect of **user activity** (measured as the proportion of *edits per user*) on growth rates. The results suggest that user activity correlates very strongly with wiki growth, not only in terms of content production (which is to a certain extent unsurprising) but also new member recruitment. The effect becomes stronger with initially more populated wikis: the more users are actively editing, the more a wiki grows in content *and* population.
- **Significant governance factors.** Turning to governance features, we first analysed the effects of the **administrator density** on wiki dynamics by examining the impact of the overall proportion of administrators per page. Figure 4.2 shows that having a relatively high number of administrators for a given content size is likely to reduce growth. There is a strong effect of the proportion of admins per page both on user and page growth. For instance, while the last quantile of admins/page ratio has near-zero growth rates over 8 months, the first quantile

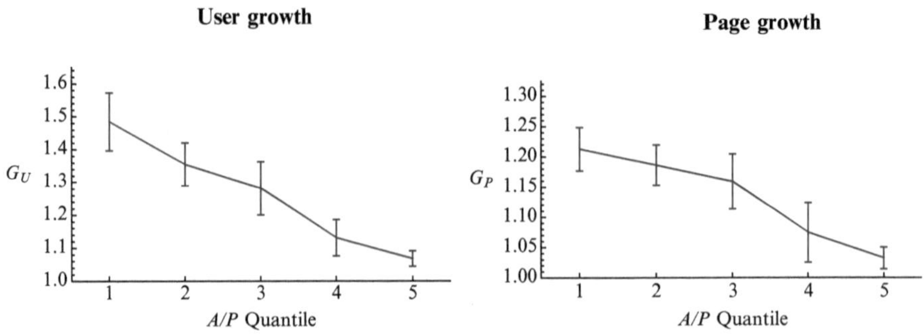

Fig. 4.2 Growth landscape with respect to the proportion of *admins density* (A/P).

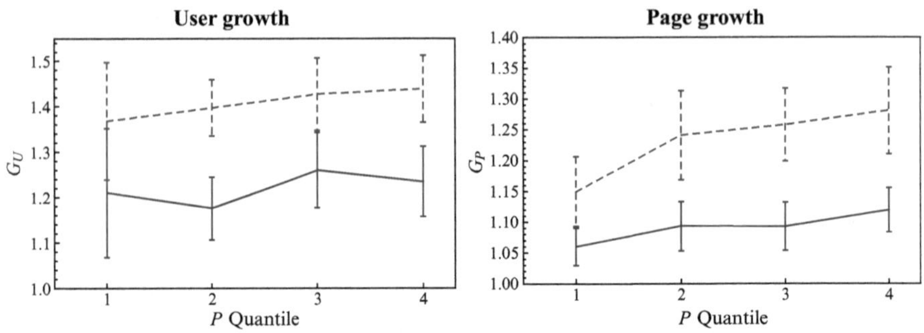

Fig. 4.3 Growth landscape with respect to *editing permission* (R): *red dashed* refers to anonymously editable wikis, while *blue solid* to wikis editable by registered users only.

shows high overall rates (∼+50% for users, ∼+25% for pages). This effect may be interpreted as the impact of strong governance activity on the proliferation of content and users.

We identified another significant effect when we considered **editing permission**. As a binary variable, the editing permission variable generates only two groups of wikis (wikis that allow anonymous editing *versus* wikis that restrict editing to registered users only). The growth landscape is consequently limited to a one-dimensional comparison over population quantiles. The results in Fig. 4.3 show that for both dimensions – population and content – having no access control is likely to favor growth. While a stronger page growth is quite unsurprising in wikis where no registration is required, the fact that this factor also fuels user registration is more puzzling. One might expect that if users can participate without the need of registration, few would be inclined to register. Our results suggest that on the contrary wikis with unrestricted registration trigger participation more easily than wikis that restrict access.

We also considered two indicators that showed a markedly milder correlation with wiki dynamics. On the one hand, we found that **edit density** (i.e. edits/page) correlates in a moderately negative way with user growth – with a relatively stronger

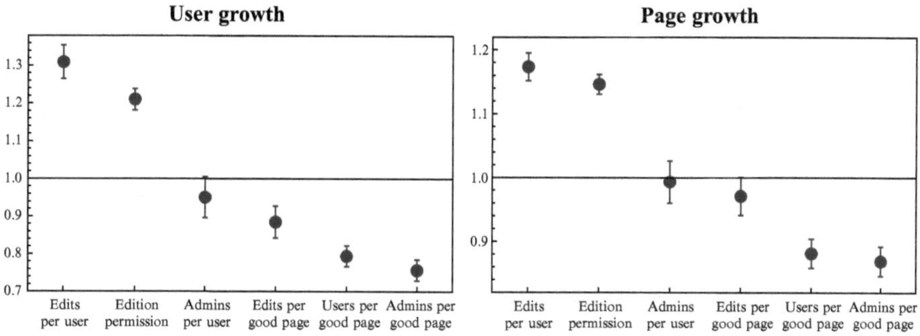

Fig. 4.4 Comparison of growth rates between last and first quantiles, for each variable.

effect depending on initial population size – while there is surprisingly no significant correlation with page growth. On the other hand, higher **administrator ratios** (i.e. admins/user) have no significant effect on content or population growth.

Figure 4.4 summarises the correlations found between growth rates and each of the variables we considered, by comparing the gain in the population and content sizes between the last and the first quantile for each variable (variables in Fig. 4.4 are ranked from the most positively to the most negatively correlated). If we focus on *structural aspects* of wikis, we note that the higher the ratio of *edits per user* the faster the wiki grows, both in terms of content and population. Wikis with very active user communities are not only likely to grow in content, but also to attract a large number of new contributors. This result contrasts with the opposite effect produced by high user density per page.

As far as *governance factors* are concerned, we observed the singular fact that population growth is in average more than 20% faster for anonymously editable wikis. This seems to support the intuition that less barriers favor population growth. Furthermore we observed that, while too many administrators per page may hinder the growth of a wiki (in terms of content size), the proportion of administrators per user does not appear to show a significant influence on growth. In all the above cases, we observed a striking correlation between content and population growth.

This approach broadly draws attention to the remarkable intertwinement of population and content growth in a relatively large sample of (wiki-based) online communities, and constitutes a first contribution towards more comprehensive research on factors behind sustainable wiki communities, beyond the dominant example of the *Wikipedia*. In particular, we endeavored at connecting simple quantitative features of these online groups to more qualitative characteristics mainly pertaining to simple organisational properties, including distribution of roles, modes of regulation and access control. In a more pragmatic perspective, it basically constitutes an overview of indicators that wiki communities should take into account in order to control their demographics, by paying specific attention to some variables and acting upon them when possible, while neglecting others.

From the perspective of viability theory, the dichotomy we propose between descriptive indicators and governance factors is meant to reflect the traditional distinction between *autonomous dynamics* and *control features*. In this respect, our results offer empirical grounds for the design of realistic models of the demographic evolution of these communities, the characterisation of their viable states and of factors that control their dynamics.

4.2.2 Social Media Communities

Flickr.com, one of the most popular photo and video sharing services, represents another ideal case for the study of online community viability, focusing here on the joint effects of content-based interaction, group affiliation and social network dynamics. The platform supports a dedicated infrastructure for the creation of communities of interest or "groups", which represents an ideal testbed for studying group viability issues, as well as, more broadly, the effect of *user-to-group* affiliation links on user behaviour and social interaction among users. The user model of Flickr additionally allows the creation of (*user-to-user*) "contact" links that can provide a direct insight into user-centered social networks; it also allows interactions among users that are mediated by content (such as commenting on a picture or marking a picture as a "favorite"), hence offering the opportunity to study social behaviour mediated by *user-to-content* links; finally, thanks to a rich and extensively documented API,[8] Flickr enables the extraction of large datasets that can be used to study social dynamics at each of these levels of description (content, users, groups).

Flickr attracted a fairly large attention in the research community. Most studies used Flickr as a large data source to study tagging behaviour and folksonomy (Marlow et al. 2006; Nov et al. 2008; Plangprasopchok and Lerman 2009; Sigurbjörnsson and van Zwol 2008). A smaller number of works, more relevant to the present analysis, focused on aspects of social interaction and group-driven behaviour (Lerman and Jones 2007; Mislove et al. 2008; Leskovec et al. 2008; Cha et al. 2008; van Zwol 2007; Valafar et al. 2009) . Reviewing the contribution of the literature on the understanding the functioning of communities in online social media is beyond the scope of this chapter. In this section we focus on the relative performance of groups at attracting members and securing content, an issue that has been addressed only in a tangential way by the above studies.

4.2.2.1 Flickr Groups

A central social feature of Flickr, i.e. *groups*, has attracted to date a modest attention in the literature, even though it is estimated that a large part of content-mediated

[8] http://flickr.com/services/api.

interactions and social interactions happens via groups. Flickr groups are of particular interest to the present analysis because, as opposed to purely user-centered social networks, they can be described as communities of interest driven by shared content. Flickr groups are specifically designed to enable *collaborative* content production and dissemination. In order to share content with the members of a group, a user is explicitly required to submit it to the group. Furthermore, groups have a governance structure consisting of at least one administrator (by default, the group creator) and an optional number of moderators. Group admins and moderators can control the rate and type of submitted content that is shared in the group, via *moderation* tools, post-submission *pruning* or *throttling* (i.e. limiting the number of posted items over a given period of time). These features make Flickr groups ideal candidates for research on content-based online social behaviour and on the coevolution of social and affiliation links. Previous research already partly addressed the role of Flickr groups from this angle (Mislove et al. 2007; Prieur et al. 2008; Schifanella et al. 2010; Negoescu and Perez 2008; Backstrom et al. 2008; Zheleva et al. 2009).

All in all these various results suggest that group formation processes in content-based communities arise from the joint effect of a large number of factors that cannot entirely account for the evolution of a group when considered on an individual basis. The question that we wish to ask is how these different factors interact in affecting the evolution in content and population of a group.

4.2.2.2 Variety of Group Dynamics

There is a striking variety in growth patterns of Flickr groups when observed over time, even if we focus on macroscopic indicators such as population and content variations: some groups are characterised by a steady population growth accompanied by a null or negative content growth (which may prima facie suggest tight moderation or regular pruning of content); other groups rapidly grow in content but vary slowly or remain virtually constant in population (suggesting the use of groups as "dumps" of pictures with little recruitment of new members); other groups show fluctuations in both content and population (suggesting a significant portion of members who leave the group when no more active); finally, groups may display sudden bursts of growth in content and population and remain subsequently inactive for long periods (which may be the case for groups about recurring or temporally discrete events).

Groups also substantially vary in member *turnover*, i.e. the portion of a group's population that is replaced by new members joining the group over time while former members leave. Some groups have a relatively low turnover, suggesting that members tend to stick in the group and are reluctant to leave, while other groups have much faster member replacement rates.

One possibility to come to grips with this variety in global dynamics might then be to ask whether groups can be broadly categorised in a qualitative way into distinct typologies, considering for instance how content specificity or content policies affect the overall group evolution over time. The alternative approach that we take in the present study consists in assuming that similarity in temporal dynamics can be traced back to group similarity in terms of structural features.

Regardless of content, each group can be characterised as occupying at a given time a region in a multidimensional space of properties defining its demographic profile, its structure and its governance mode. These properties can pertain to a group as a whole or refer to aggregate properties of its members, such as their average degree or group affiliation spread.[9] The temporal dynamics of a group can then be studied as a trajectory across this space. Our study aims to find regularities in the observed temporal dynamics of a large set of groups by assuming that a number of initial properties of these groups can be explored as predictors of their macroscopic evolution – which, again, is taken as a preliminary description of some dimensions of their viability. The literature on group affiliation dynamics offers a number of suggestions as to how groups are generally expected to evolve over time as a function of their size, structure and properties of their membership:

- P1 **Larger groups tend to grow faster** than smaller groups, in virtue of a preferential attachment principle.
- P2 **Cohesive groups tend to recruit less new members** than weakly cohesive groups, because of a stronger social closure (or "cliquishness"), which also results in an increased membership inertia and less user turnover.
- P3 **Groups whose members are sociable tend to grow faster** and attract more contributions than groups whose members have a relatively small number of friends.
- P4 **Highly curated groups tend to grow slower in content but faster in population** because of the competitiveness produced by higher content selectivity.
- P5 **Groups whose members belong to many other groups grow less in content** than groups with members that belong just to a few groups.

Each of these hypotheses can be empirically explored, by considering the observed growth rates over a specific time frame as a function of characteristic properties of a group.

4.2.2.3 Dataset

The data used for this study consists of a sample of 9,360 *public* Flickr groups whose variations were tracked on a daily basis for a period of 1 month between June and July 2009. The data was obtained via *Flickr Group Trackr*[10], a public Web service that we developed in order to allow Flickr group members to track the daily evolution of their community. For each group registered to the service, *Flickr Group Trackr* pulls a series of statistics from the Flickr API on a daily basis, including: size

[9]For a comparison with the idea of a sociodemographic space, see McPherson et al. (1992).
[10]http://dev.nitens.org/flickr/group_trackr.php.

of the group pool (or number of pictures uploaded to the group), population, privacy level, moderation properties, throttling type and level. Changes along any of these variables can hence be identified with a precision of 24 h. It should be noted that we did not consider group activity data related to discussions in group forums as this data are not available via the Flickr API. The dataset thus obtained from *Flickr Group Trackr* was complemented with a static snapshot of the same set of groups providing data on: *user-to-group* affiliation links and *user-to-user* contact links.

The dataset was filtered in a number of ways to obtain a more homogeneous sample. We limited our analysis to a set of medium-to-large groups with a population range of 100 to 100,000 members; this restriction was introduced to avoid biases in the analysis due to the presence of small groups ($u_0 < 100$), whose dynamics are too dependent on the behaviour of individual members to allow any useful generalisation.[11] To capture the natural dynamics of these groups we also introduced a capping on the maximum daily growth rate in content and population, as we did for wikis, excluding those groups displaying an instantaneous growth of more than 5% of their pool size or population (possibly resulting, again, from extrinsic events such as contests or administrator bulk decisions). Groups that switched to *private* access control mode during the tracking period were also excluded from the sample. As a result of these restrictions, the final dataset used here consists of 9, 167 groups.

The dataset also contains a complete snapshots of the population of each of the tracked groups at t_0 as well as the complete list of contacts and affiliations for each member of these groups. The union of members of the groups in the dataset spans a total population of 1,267,874 unique users. Group pools size and group populations in our dataset follow a log-normal distribution.[12]

4.2.2.4 Variables

The metrics that we used as independent variables to study the drivers of group dynamics throughout the present study are described in Table 4.3. Among demographic metrics, ms (membership spread) indicates the number of other groups a group member is affiliated with, averaged over the whole group population. Among structural metrics or metrics related to topological properties of the group social network: k refers to the direct degree for group members calculated on contact

[11] A similar rationale for focussing on mid-size groups is in Backstrom et al. (2006).

[12] It should be noted that, compared to other studies that considered a random sample of the global Flickr user and group population, our study focused on *public groups* (i.e. groups with public content, flagged as "safe" and hence open to recruit any Flickr user as a potential member) and users that engage in actual social activity such as being member of at least one public group (Prieur et al. 2008 estimated this to represent the 8% of the total Flickr population in 2006). This explains the mismatch between the global statistics reported by other studies that include private groups and non-social users (i.e. users who may only use social media services as a way to dump private content not meant for public consumption).

Table 4.3 Flickr group metrics used as independent variables

	u_0	Number of group members at time t_0
	p_0	Number of photos in the group pool at time t_0
Demographic metrics	ms	Average membership spread of group members
	k	Average directed degree of group members
	c_3	Average clustering coefficient
Structural metrics	r	Reciprocity index
	adm	Number of group administrators
	mod	Number of group moderators
	μ	Moderation filter
Governance metrics	θ	Throttling index

Table 4.4 Flickr group growth indicators

	Δu	Absolute population variation ($u_1 - u_0$)
Group size variation	Δp	Absolute content variation ($p_1 - p_0$)
	u_+	Number of users who joined the group
Population turnover	u_-	Number of users who left the group

links that are internal to the group social network; r measures the proportion of *reciprocated* or symmetrical contact links within the group and per group member, averaged over the group population. Among governance metrics: mod indicates the number of superusers other than administrators who can accept photos submitted to the group's moderation queue; μ indicates the presence of a moderation queue, by which photos submitted to a group are reviewed by moderators before being published in the group pool; θ is a quantitative indicator of the maximum number of photos that can be contributed to the group per time period (day, week or month), also denoted as "throttling index".

Group growth indicators can be defined in multiple ways. Growth can be assessed in absolute terms as the difference in the total number of members and photos between t_0 and t_1, i.e.: $u_1 - u_0$ and $p_1 - p_0$, respectively. Alternatively, one may focus on relative growth or "growth rate" over the observation period, or the variation in members and content normalised by the initial size of the group: $\frac{u_1-u_0}{u_0}$ and $\frac{p_1-p_0}{p_0}$. Finally, one may consider the actual turnover or the number of unique users who joined (u_+) and leaved a group (u_-) over the observation period. The turnover itself can be considered in absolute ($u_+ - u_-$) or relative terms $\frac{u_+-u_-}{u_0}$.

For the sake of the present study and contrarily to the wiki case, we decided to focus on absolute rather than relative growth indicators (see Table 4.4) for a number of reasons. First of all, we wanted to take all groups at face value as equally prone to recruit new members and measure size-dependent effects as only one among several possible assumptions on growth driving factors. Although other studies showed that the size of a group population plays a central role in the recruitment of new members (Backstrom et al. 2008), this assumption can be challenged on the basis of the significant number of group members who do not appear to have any social connection with other members ($k = 0$). Evidence of the existence of such members, as pointed out by Zheleva et al. (2009), suggests that "social recruitment" is only one among possible mechanisms that attract new members to a group. Second, we wanted to study the specificity of member turnover as indicators of a stable or volatile community, and for this reason we also decided to opt for absolute figures as opposed to relative growth rates. A final reason not to focus on relative growth rates was that results using these rates as dependent variables were not statistically significant in several cases, suggesting that for the timeframe that we considered absolute variations were the most appropriate to focus on.

4.2.2.5 Aggregate Analysis of Growth-Driving Factors

To investigate the joint contribution of demographic, structural and governance-related factors on the temporal dynamics of groups, we performed a regression analysis of absolute group growth over the whole observation period as a dependent variable. We used four different models aiming at measuring the respective effect of a series of independent variables on absolute user variation (Δu), content variation (Δp) as well as member turnover (u_+ and u_- respectively). We used the initial population and content size as control variables in each of the models (see Table 4.5 for the detailed list of variables included in each model). The general regression equation underlying each model (barring specific variable exclusions) is:

$$\begin{aligned}\log(y) = {} & \lambda_0 + \lambda_{u_0}\log(u_0) + \lambda_{p_0}\log(p_0) \\ & + \lambda_r(r) + \lambda_{c_3}(c_3) + \lambda_k\log(1+k) + \lambda_{ms}\log(ms) \\ & + \lambda_\mu\mu + \lambda_{mod}\log(1{+}mod) + \lambda_\theta\log(\theta) + \lambda_{adm/u0}\log(1+(adm/u_0))\end{aligned} \tag{4.1}$$

We thus considered a linear regression of the logs of each variable, when applicable and relevant: logs were essentially used for quantitative variables spanning over one or several orders of magnitude (such as u_0) in order to make them comparable in the regression with variables evolving in e.g. $[0, 1]$ (such as c_3). For each dependent variable $y \in \{\Delta u, \Delta p, u_+, u-\}$, we started with an equation specified by the full model of (4.1). Variables corresponding to non-significant p-values were then iteratively excluded, generally resulting in a change in R^2 of less than 1%.

The results of the regression analysis summarised in Table 4.5 indicate some salient effects of various initial properties of groups on their dynamics. For a given dependent variable, an empty cell indicates that the corresponding independent variable had eventually been excluded from the regression. If we focus on **population and content growth**, we first notice a (somewhat unsurprising) correlation in the effect of different variables on population growth on the one hand and content growth on the other hand, which is consistent with the above findings in wiki-based communities. As to structural/demographic factors, we observe indeed that population (u_0) and pool size (p_0) are important drivers of absolute growth: the larger the population of a group, the stronger its absolute growth over the

Table 4.5 Results of regression analysis

	Population variation (Δu)		Content variation (Δp)		Population turnover			
					Joining (u_+)		Leaving (u_-)	
Parameter	Value	p	Value	p	Value	p	Value	p
λ_{u_0}	0.87	***			0.77	***	0.78	***
λ_{p_0}			0.94	***	0.11	***	−0.03	***
λ_r	0.99	***	2.09	***			−0.19	**
λ_{c_3}	−1.87	***	−1.73	***	−1.49	***	−1.27	***
λ_k	0.10	*	0.18	***			0.23	***
λ_{ms}	−0.57	***	−0.33	***	−0.43	***	0.35	***
λ_μ					0.08	**	0.07	***
λ_{mod}	0.05	**	0.09	***	0.08	***	0.02	***
λ_θ			0.10	***				
λ_{adm/u_0}					−0.06	***		
R^2		*0.65*		*0.75*		*0.68*		*0.82*

Log-linear regression model on absolute population variation (Δu), absolute content variation (Δp) *(left)* and population turnover measured as absolute number of joining members (u_+) and leaving members (u_-), respectively *(right)*. Significance: *, ** and *** mean a p-value smaller than 0.05, 0.01 and 0.001 respectively

observation period (consistently with **P1**). The average spread of group affiliation for group members (ms) displays a negative correlation, suggesting that groups whose members also belong to many other groups tend to grow slower and the effect is actually stronger on population growth than it is on content (**P5.**): this is consistent with the idea that groups whose members are selective (i.e. choose to join a smaller number of groups) are likely to attract more members than groups that mostly function like content dumps for occasional members. In terms of topological properties of the group-centered social network, we observe that cohesiveness as measured by the average clustering coefficient of the group-based network (c_3) has a remarkable negative correlation with growth (**P2**) and is by far the variable displaying the strongest effect across all analyses. Conversely, a high rate of reciprocity (r) and a larger presence of (popular) high-degree nodes in a group (k) have the effect of boosting growth (**P3**). Possibly the most striking finding is the overall negligible effect of moderation properties on the observed growth. In many cases the effect of moderation factors ($\mu, mod, \theta, adm/u_0$) is not statistically significant; in those cases in which it is, the observed effects are considerably weaker than those related to other group properties, which is partly at odds with our expectations (**P4**).

The analysis of factors affecting **member turnover** provides further insights. The strongest effect on turnover is that of cohesiveness, which not only appears to hinder new recruits but also to work as a barrier against user drop-off, as indicated by its negative effect on both components of the turnover: as such, cohesiveness (or the "cliquishness" of a group) works as a factor measuring the social inertia of a group membership, suggesting a higher level of commitment by its members that are more reluctant to leave than in less cohesive groups (**P2**). The level of engagement is also measured by the symmetric effect that affiliation spread has with respect to member recruitment and drop-off: a higher spread increases the probability that more members will be leaving the group and less new members joining.

4.2.2.6 Individual Drivers of Group Growth

Whereas the regression results can be used for a global assessment of the contribution of different factors to the dynamics of a group, we can address each of the hypotheses presented in section 4.2.2.2, following the methodology adopted before for wiki communities. We tackled the implications of this regression model on each hypothesis through an analysis of the individual impact of each metric on the observed growth and turnover of a group. Two snapshots for each group were compared at the beginning (t_0) and at the end (t_1) of the tracking period and group growth rates were calculated as the absolute variation in population and pool size between these two snapshots (Δu and Δp respectively). We then ranked groups along each independent variable in 9 quantiles, each containing therefore $1/9$ of the groups in our dataset. The first quantile represents groups with the lowest values for the considered variable, whereas the last quantile refers to groups with the highest values. The analysis of individual effects should be taken as evidence of how effective each factor would be under the assumption that all other factors had an equal effect on growth.

- **P1: Size matters.** The breakdown of the effects of size on the observed growth (Fig. 4.5) shows indeed that the expected growth of a group in content and population follows monotonically from its size. This allows us to discard the null assumption that we made that all groups should in principle be considered at face value as having an equal probability of attracting new members and new content: size does matter, which can be explained by e.g. herding behaviour.
- **P2: Effects of cohesiveness on group growth.** Figure 4.6 (*left*) shows the breakdown of the effects of cohesiveness on group growth. Consistently with

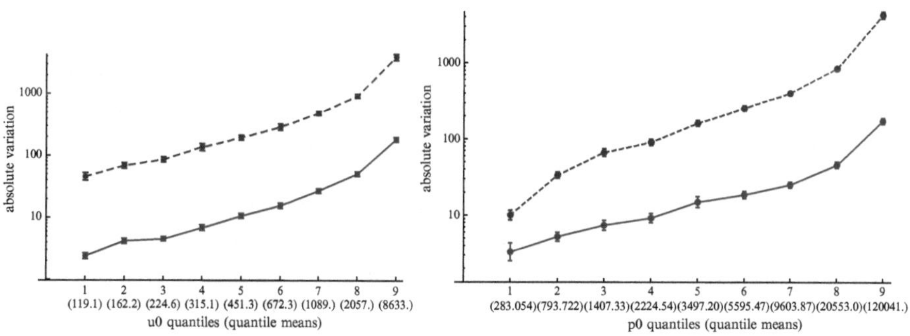

Fig. 4.5 **P1**: Size matters. Effects of u_0 and p_0 on absolute variation of users (*solid line*) and photos (*dashed line*)

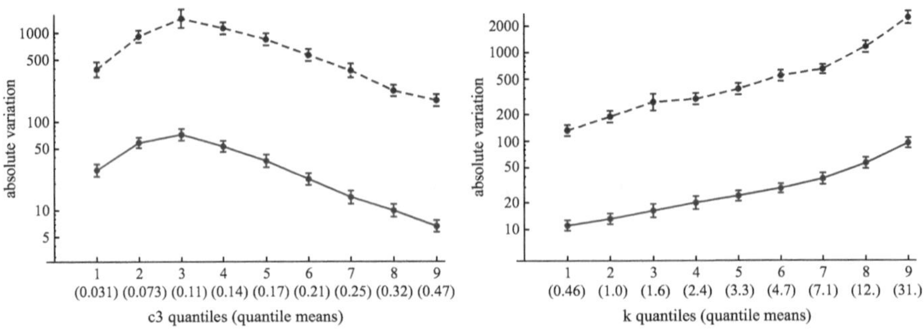

Fig. 4.6 (*left*) **P2**: Effects of cohesiveness (c_3), (*right*) **P3**: Effects of sociability on absolute variation of users (*solid line*) and photos (*dashed line*) (k)

the regression analyses, cohesiveness as measured by the average clustering coefficient for the group-centered network works as a growth-regulating factor. Groups where cohesiveness is high display a higher inertia.

- **P3: Are sociable users growth attractors?** An effect conflicting with cohesiveness is related to individual sociability as measured by the average *within-group* degree of members in the group-centered contact social network (Fig. 4.6, *right*). Note that degree has only been measured within groups: a related hypothesis, assessing sociability through a degree computed over the whole network of the Flickr population would actually allow one to answer the question whether groups in which (global) high-degree nodes (or very social/popular users) are concentrated are more likely to attract members than groups where the degree is more uniformly distributed.
- **P4: The poor effects of governance.** Possibly the most striking findings of the present study are the negligible effects of the moderation and governance structure on group growth. Figure 4.7 (*left*) exemplifies the virtually flat growth landscape that emerges as a function of θ. This is not to deny the effectiveness of curators' strategies in actually enforcing norms about content and participation on group members. However, from a purely quantitative perspective, these results suggest that in social media sharing systems, social-network factors are likely to drive recruitment and participation to a much larger extent than the factors that group administrators and moderators can control with the help of governance tools. This result contrasts with the above findings on wikis, which raises the question of what differences in terms of user interaction modes and collaborative behaviour may explain this discrepancy.
- **P5: User engagement and attention.** The marginal role of governance-related factors suggests that the main drivers of group dynamics in social media sharing systems need to be found elsewhere. In addition to social ties, individual and collective attentional spans may influence group growth. We saw that affiliation spread (ms) has a globally significant effect on group growth; analysing growth as a function of different values of affiliation spread (Fig. 4.7, *right*) indicates that this effect is robust also at an individual basis: groups whose members tend

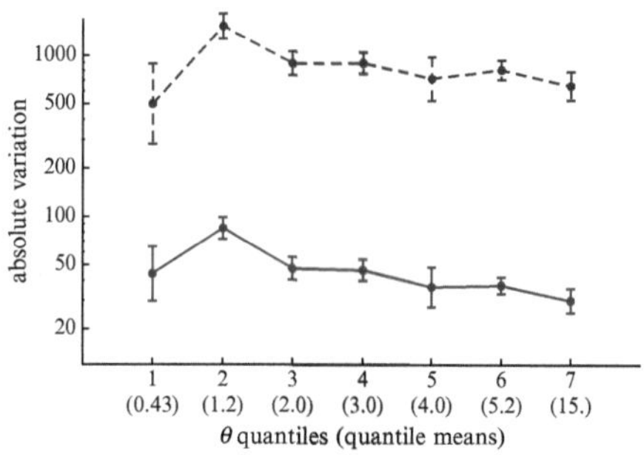

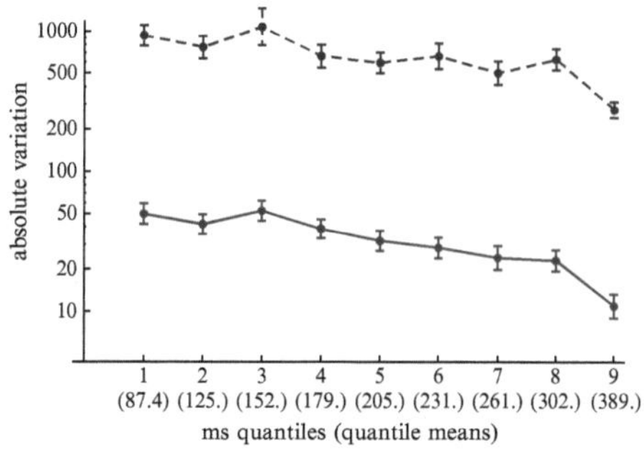

Fig. 4.7 *(left)* **P4**: Weak effect of governance, exemplified by throttling (θ). *(right)* **P5**: Level of user engagement measured by affiliation spread (ms). Plots represent absolute variation of users *(solid line)* and photos *(dashed line)*

to spread their contribution over many other groups are consistently slower in growth than groups whose members are more selective in their affiliations.

From a viability theory perspective, these results diverge from the relatively clear-cut distinction between descriptive factors and governance factors which we discussed in the case of wikis. Indeed, while it would not be hard to consider plain demographic metrics such as population and content size as characterising of the autonomous dynamics of a model of social media groups, typical governance measures (such as throttling or moderation), appear to have little influence on the evolution of these groups. In this regard the topological structure of group-centred social networks exerts a more important effect, however indirect. As noted above, although these factors can be seen as defining the autonomous dynamics of the system (by assuming that social ties are the result of spontaneous social interaction among users) they can also be regarded as genuine control factors (insofar as the design of the system or the implementation of specific policies can favour the creation of some kind of links, e.g. transitive links, or increase the cohesiveness of the members' social network).

4.3 A Simple Model of Viable Web Communities

In this section we propose to formulate a simple model in which the viability of collaborative communities could be assessed against constraints on group population and group content size. In other words, we sketch the initial steps needed to formally account for the intertwinement of these constraints, the autonomous demographic dynamics of such systems and the possible control actions which may be adopted to ensure and/or restore their viability. We leave issues specifically related to model solving and simulation-based approximation outside of this chapter. To this end, the interested reader may nonetheless apply the tools and techniques presented in Part 2 of the present book.

4.3.1 Model Variables

From the two examples discussed in the previous section we may abstract a list of variables that can be applied to describe the structure and governance of collaborative systems in general. Apart from the fundamental demographic metrics consisting of population and content sizes, such communities are generally characterised by the fact that some users (so-called "administrators") have special privileges; besides, they often feature similar governance mechanisms that limit the amount of contributions (filtering processes, registration requirements before contribution, etc.).

Model variables are to be based upon the corresponding quantities, at least for those features that can be described in quantitative terms. Among these variables, we have to distinguish *state variables* from *control variables* (upon which, for instance, administrators may act in order to influence the dynamics of the community they manage). Additionally, we must be able to define a viability domain, wherein we expect state variables should remain in order to consider the system viable.

4.3.1.1 State Variables

The state space per se is essentially made of:

- U**:** denoting the population of members or participants in a community.
- P**:** representing the size of content contributed by members (pages, photos, etc.).

4.3.1.2 Control Variables

Those include:

- A**:** the number of administrators in a community. For convenience, we also call *administrator ratio* the proportion of administrators with respect to the whole population, denoted as a; similarly, the *administrator density* is defined as the proportion of administrators with respect to content size, denoted as b. We then have: $aU = A$ and $bP = A$.
- m**:** moderation constraint. The moderation constraint represents a mechanism that filters contributions, such as: e.g. edition permission on wikis ($m = R$), or moderation queues in Flickr ($m = \mu$).

Eventually, the status of:

- c**:** cohesiveness (for instance measured through the clustering coefficient: $c = c_3$ in Flickr groups)

as a control variable remains relatively unclear: on the one hand, it may indeed be possible for administrators to favour some kinds of interactions between users, in such a way that the group becomes more cohesive. Yet, it is likely that the evolution of this variable could also be dictated by the autonomous dynamics of the group – for instance and all other things being equal, an increasing population is likely to induce a weaker cohesiveness.

Viability domain

As discussed in the first section, the definition of a "desirable state" as a precondition to studying group viability is bound to have a large number of potential interpretations in social systems or, at least, be more debatable than in the case of e.g. physical systems. In the following preliminary model, we choose to adopt a rather simple approach to viability by stylising and extrapolating a plausible trend observed both in Flickr and wiki groups: that groups tend to roughly grow along a diagonal of constant ratio U/P. In particular, we notice that the largest groups (both in terms of content and population) are concentrated along this diagonal. We suggest that these groups should have had a successful development, at least at some point, in order to reach this area of the (U, P) state space (Table 4.6).

We thus define viable a group such that its U/P ratio remains within a given boundary:

$$\rho = [\rho^-, \rho^+]$$

4.3.2 *Viable Dynamics*

The model aims at stylising several effects initially observed in the empirical data. In its present version however, it values mathematical tractability over realism. From a generic point of view, the system of differential equations governing the evolution

Table 4.6 Model variables and parameters

	Name	Description	Range
	U	Group population	$\mathbb{R}^{+*}$
	P	Group content size	$\mathbb{R}^{+*}$
	A	Administrator population	$\mathbb{R}^{+*}$
	m	Moderation constraint	$\{0,1\}$
	c	Cohesiveness of contacts within the group	$[0,1]$
Variables	r	Proportion of reciprocated contacts	$[0,1]$
	λ_a	Slope of growth vs. Population size	$\mathbb{R}^{+*}$
	λ_b	Slope of growth vs. content size	$\mathbb{R}^{+*}$
	ν_a	Relative spread of administrator ratios (per user)	$[0, 1]$
	ν_b	Relative spread of administrator densities (per page)	$[0, 1]$
	g_0, g_1	Effect of resp. the absence or presence of moderation	$\mathbb{R}^{+*}$
Parameters	ρ^-, ρ^+	Minimum and maximum population/content ratio	$\mathbb{R}^{+*}$

of state variables could be written as:[13]

$$\begin{cases} \dfrac{dU_t}{dt} = \alpha(U_t, P_t, A_t, c_t, m_t, r_t) \\ \\ \dfrac{dP_t}{dt} = \beta(U_t, P_t, A_t, c_t, m_t, r_t). \end{cases} \tag{4.2}$$

From this point, various equations and diverse models can be proposed depending on different collaborative models one wants to describe, peer production systems (such as wikis) vs. social media groups (such as Flickr communities). We choose to focus on wikis and, in particular, their user dynamics, dU_t/dt. In the absence of empirical results based on social network properties in wikis we should ignore the role of cohesiveness in the present model. Eventually, we can write:

$$dU_t/dt = \alpha_{\mathrm{W}}(U_t, P_t, A_t, m_t)$$

In our previous analyses, we made the hypothesis that wiki growth over the *whole* observation period was directly influenced by a constant and permanent impact of the values of control and state variables measured at the *beginning* of the given period. The most straightforward way to account for this type of growth is to assume that groups experience an exponential growth depending on initial conditions: in other words, given U_{t_0}, P_{t_0}, A_{t_0} and m_{t_0}, growth is assumed to be an exponential function of t.

The following equation can therefore be proposed, for a proper function ϕ:

$$\frac{dU_t}{U_t} = \phi(U_{t_0}, P_{t_0}, A_{t_0}, m_{t_0})dt \quad \Leftrightarrow \quad U_t = U_{t_0} e^{(t-t_0)\phi(U_{t_0}, P_{t_0}, A_{t_0}, m_{t_0})}. \tag{4.3a}$$

where ϕ could be estimated for instance from Roth et al. (2008) as a function depending (1) linearly upon U_{t_0}, (2) affinely and in a monotonously decreasing manner upon A_{t_0}/P_{t_0} (or, rather, quantiles thereof) and (3) upon a given step function of m_{t_0}. Put differently, ϕ can be schematically written as:

$$\begin{aligned} \phi(U_{t_0}, A_{t_0}, P_{t_0}, m_{t_0}) &= \lambda_a U_{t_0} \left(\frac{b_{t_0} - b_{\max}}{b_{\min} - b_{\max}} \right) g_{m_{t_0}} \\ &= \frac{\lambda_a}{\nu_b} U_{t_0} (1 - \frac{A_{t_0}}{b_{\max} P_{t_0}}) g_{m_{t_0}} \end{aligned} \tag{4.4}$$

[13]For the sake of differentiability, we obviously have to make a continuous approximation on U, P and A by considering that they evolve in $\mathbb{R}$; one may still consider that the population at t equals $\lfloor U_t \rfloor$.

where $\lambda_a \in \mathbb{R}^+$ is a given constant, $b_{\min}$ and $b_{\max}$ are respectively the minimum and maximum administrator densities over all groups, $\nu_b = \frac{b_{\max}-b_{\min}}{b_{\max}}$ is the constant relative spread of administrator ratios, and g_0 and g_1 are two given constants such that $g_0 < g_1$ (in practice, from Fig. 4.3 it roughly seems that $g_1/g_0 \in [2,3]$).

4.3.2.1 Viable Group Dynamics: A Coevolutionary Sketch

By extrapolating on this empirically-based relationship, we propose a dynamic model featuring the following system of equations:

$$\begin{cases} \dfrac{dU_t}{dt} = \dfrac{\lambda_a}{\nu_b} g_{m_t} (1 - \dfrac{A_t}{b_{\max} P_t}) {U_t}^2 \\ \dfrac{dP_t}{dt} = \dfrac{\lambda_b}{\nu_b} g_{m_t} (P_t^2 - \dfrac{A_t P_t}{b_{\max}}) \end{cases} \tag{4.5}$$

while $\forall t \geq 0$, the following double inequality should hold:

$$\rho^- < U_t/P_t < \rho^+.$$

Possible refinements, given specific cases and under specific subsets of assumptions, where for instance A_t itself could be connected to U_t and P_t, are left to the reader. The same goes for the viability domain: to account for the need for a group to keep receiving contributions, constraints should be put onto dP_t/dt – for instance, by requiring that it remains above a certain level of activity and below a certain threshold of cognitive effort for group members (so that they can reasonably keep track of ongoing contributions). Similarly, exact solutions of this class of models are left to the interest of the reader. Approximate and simulation-based exploitation of these kinds of dynamics can be exploited through software such as Kaviar, to allow performing computations of the corresponding viability kernels (see Chap. 7).

4.4 Conclusions

This chapter presented a methodology aimed at empirically appraising one possible dimension and understanding of "viability" (growth-related viability), in two paradigmatic cases of collaborative communities. In this respect, it constitutes a preliminary framework and a necessary step towards defining and modelling pattern resilience and viability in the context of social systems.

In particular, we assessed the interplay of demographic factors and governance structure and their role among forces driving the macro-level growth of content-based Web communities; in the case of social media communities, we emphasised the role of social network properties as a class of indirect control factors. Despite a largely shared ontology (governance features, demographic factors) between the two case studies and the application of a similar analysis taking growth as a proxy for viability, we noticed several macroscopic discrepancies: most importantly, we observed that demographic and governance properties are good predictors of growth in the case of wikis, but are surprisingly poor at predicting growth rates in social media communities. Some possible reasons for this discrepancy may depend on the specific methodology that we adopted here (e.g. unnoticeable effects on relative growth rates as opposed to absolute growth). The observed inconsistency in the effectiveness of control factors may be due to distinct underlying modes of peer production–i.e. genuinely *collaborative*, in the case of wikis (as users jointly modify shared content) vs. *atomistic* in the case of media sharing (as users contribute individual contents to a shared pool) – or distinct types of social interaction (suggesting the effects of the underlying social network as a stronger recruitment factor in social media groups).

All in all, we characterised the relationships between structural, demographic and governance-related variables of these online groups. We interpreted these relationships from the perspective of the evolution, stability and sustainability of these Web communities. We essentially showed how data can be analysed and how, in principle, the results of such analysis could be used as input for highly aggregated mathematical models: within the framework of viability theory, we argue that this approach makes it possible to define hypotheses pertaining to the *autonomous dynamics* of such systems and the factors that control them, and therefore serve as a first step towards the design and computation of the corresponding "capture basins" or regions of the variable space where viability can be ensured. More broadly, we suggest that it constitutes a preliminary framework and a necessary step towards defining and modelling pattern resilience and viability in the context of social systems.

Acknowledgements This work was partly supported by the PATRES project (NEST-043268) funded by the FP6 programme of the European Commission and by the Future and Emerging Technologies programme FP7-COSI-ICT of the European Commission through project QLectives (grant 231200). We are grateful to Nigel Gilbert, Volker Grimm, Nic Geard, Przemek Grabowicz, as well as members of the Centre for Research in Social Simulation (University of Surrey) and participants in the 2008 Dagstuhl seminar "Social Web Communities" for valuable feedback and insights on earlier versions of this work.

References

Almeida R, Mozafari B, Cho J (2007) On the evolution of Wikipedia. In: Proceedings of the international conference weblogs and social media (ICWSM '07)

Anthony D, Smith S, Williamson T (2005) Explaining quality in internet collective goods: Zealots and good samaritans in the case of Wikipedia. URL http://web.mit.edu/iandeseminar/Papers/Fall2005/anthony.pdf

Backstrom L, Huttenlocher D, Kleinberg J, Lan X (2006) Group formation in large social networks: Membership, growth, and evolution. In: Proceedings of the 12th ACM SIGKDD international confernce on knowledge discovery and data mining. ACM, New York, pp 44–54

Backstrom L, Kumar R, Marlow C, Novak J, Tomkins A (2008) Preferential behavior in online groups. In: Proceedings of the international conference on web search and web data mining (WSDM '08). ACM, New York, pp 117–128

Benkler Y (2002) Coase's penguin, or Linux and the nature of the firm. Yale Law J 112(3):367–445

Benkler Y (2006) The wealth of networks: How social production transforms markets and freedom. Yale University Press, London

Brandes U, Lerner J (2008) Visual analysis of controversy in user-generated encyclopedias star. Inform Vis 7:34–48

Bryant SL, Forte A, Bruckman A (2005) Becoming wikipedian: Transformation of participation in a collaborative online encyclopedia. In: Group'05, Sanibel Island, FL, USA

Capocci A, Servedio VDP, Colaiori F, Buriol LS, Donato D, Leonardi S, Caldarelli G (2006) Preferential attachment in the growth of social networks: The internet encyclopedia Wikipedia. Phys Rev E 74(3):036116

Cha M, Mislove A, Adams B, Gummadi KP (2008) Characterizing social cascades in Flickr. In: Proceedings of the 1st ACM SIGCOMM workshop on social Networks (WOSN'08). ACM, New York

Ciffolilli A (2003) Phantom authority, self-selective recruitment and retention of members in virtual communities: The case of Wikipedia. First Monday 8(12)

Edling CR (2002) Mathematics in sociology. Ann Rev Sociol 28:197–220

Forte A, Larco V, Bruckman A (2009) Decentralization in Wikipedia governance. J Manage Inform Syst 26(1):49–72

Giles J (2005) Internet encyclopaedias go head to head. Nature 438(7070):900–901

Godfrey M, Tu Q (2001) Growth, evolution, and structural change in open source software. In: Proceedings of the 4th international workshop on principles of software evolution (IWPSE '01). ACM, New York, pp 103–106

Godfrey MW, Tu Q (2000) Evolution in open source software: A case study. ICSM 00:131

Halim F, Yongzheng W, Yap R (2009) Wiki credibility enhancement. In: WikiSym '09: Proceedings of the 5th international symposium on Wikis and open collaboration. ACM, New York, pp 1–4

Happel H-J, Treitz M (2008) Proliferation in enterprise wikis. In: Proceedings of the 8th international conference on the design of cooperative systems (COOP 08). Carry-le-Rouet, France

Hars A, Ou S (2002) Working for free? motivations for participating in open-source projects. Int J Electron Commer 6(3):25–39

Kittur A, Chi E, Pendleton BA, Suh B, Mytkowicz T (2007) Power of the few vs. wisdom of the crowd: Wikipedia and the rise of the bourgeoisie. In: ALT CHI 2007. San Jose, CA

Kittur A, Pendleton B, Kraut RE (2009) Herding the cats: The influence of groups in coordinating peer production. In: WikiSym '09: Proceedings of the 5th international symposium on Wikis and open collaboration. ACM, New York, pp 1–9

Lam SK, Riedl J (2009) Is Wikipedia growing a longer tail? In: GROUP '09: Proceedings of the ACM 2009 international conference on supporting group work. ACM, New York, pp 105–114

Lerman K, Jones L (2007) Social browsing on Flickr. In: Proceedings of the international conference on weblogs and social media (ICWSM '07)

Leskovec J, Backstrom L, Kumar R, Tomkins A (2008) Microscopic evolution of social networks. In: Proceedings of the 14th ACM SIGKDD international conference on knowledge discovery and data mining (KDD '08). ACM, New York, pp 462–470

Levrel J (2006) Wikipédia, un dispositif médiatique de publics participants. Réseaux 24(138): 185–218

Lih A (2004) Wikipedia as participatory journalism: Reliable sources? Metrics for evaluating collaborative media as a news resource. In: 5th international symposium on online journalism. Austin, TX, USA

Mader S (2007) WikiPatterns. A practical guide to improving productivity and collaboration in your organization. Wiley, Indianapolis

Marlow C, Naaman M, Boyd D, Davis M (2006) HT06, tagging paper, taxonomy, Flickr, academic article, to read. In: HYPERTEXT '06: Proceedings of the seventeenth conference on Hypertext and hypermedia. ACM, New York, pp 31–40

McPherson JM, Popielarz PA, Drobnic S (1992) Social networks and organizational dynamics. Am. Sociol. Rev. 57(2):153–170

McPherson M (1983) An ecology of affiliation. Am. Sociol. Rev. 48(4):519–532

Mislove A, Koppula HS, Gummadi KP, Druschel P, Bhattacharjee B (2008) Growth of the Flickr social network. In: WOSP '08: Proceedings of the 1st workshop on Online social networks, ACM, New York, pp 25–30

Mislove A, Marcon M, Gummadi KP, Druschel P, Bhattacharjee B (2007) Measurement and analysis of online social networks. In: IMC '07: Proceedings of the 7th ACM SIGCOMM conference on internet measurement. ACM, New York, pp 29–42

Negoescu RA, Perez DG (2008) Analyzing Flickr groups. In: CIVR '08: Proceedings of the 2008 international conference on content-based image and video retrieval. ACM, New York, pp 417–426

Newman MEJ (2006) Modularity and community structure in networks. Proc Nat Acad Sci 103(23):8577–8582

Nonnecke B, Andrews D, Preece J (2006) Non-public and public online community participation: Needs, attitudes and behavior. Electron Commer Res 6(1):7–20

Nov O, Naaman M, Ye C (2008) What drives content tagging: the case of photos on Flickr. In: CHI '08: Proceeding of the twenty-sixth annual SIGCHI conference on human factors in computing systems. ACM, New York, pp 1097–1100

Plangprasopchok A, Lerman K (2009) Constructing folksonomies from user-specified relations on flickr. In: WWW '09: Proceedings 18th international conference on world wide web. ACM, New York, pp 781–790

Prieur C, Cardon D, Beuscart J-S, Pissard N, Pons P (2008) The strength of weak cooperation: A case study on Flickr. arXiv:0802.2317v1

Reagle JM (2007) Do as I do: leadership in the Wikipedia. In: Proceedings of WikiSym '07, 3rd international symposium on Wikis. ACM, New York

Roth C (2007) Viable wikis: struggle for life in the wikisphere. In: WikiSym '07: Proceedings of the 2007 international symposium on Wikis. ACM, New York, pp 119–124

Roth C, Taraborelli D, Gilbert N (2008) Measuring wiki viability. An empirical assessment of the social dynamics of a large sample of wikis. In: WikiSym '08: Proceedigns of the 4th international symposium on Wikis. ACM, New York

Schifanella R, Barrat A, Cattuto C, Markines B, Menczer F (2010) Folks in folksonomies: Social link prediction from shared metadata. In: Proceedings of the 3rd ACM international conference on web search and data mining (WSDM'10). ACM, New York

Sigurbjörnsson B, van Zwol R (2008) Flickr tag recommendation based on collective knowledge. In: Proceedings of the 17th international conference on world wide web (WWW '08). ACM, New York, pp 327–336

Stvilia B, Twidale MB, Smith LC, Gasser L (2008) Information quality work organization in Wikipedia. J Am Soc Inform Sci Technol 59(6):983–1001

Suh B, Chi EH, Kittur A, Pendleton BA (2008) Lifting the veil: Improving accountability and social transparency in wikipedia with wikidashboard. In: Proceedings of the 26th SIGCHI conference on Human factors in computing systems (CHI '08). ACM, New York, pp 1037–1040

Valafar M, Rejaie R, Willinger W (2009) Beyond friendship graphs: A study of user interactions in Flickr. In: Proceedings of the 12th ACM SIGKDD on Online social networks (WOSN '09). ACM, New York, pp 25–30

van Zwol R (2007) Flickr: Who is looking? In: IEEE/WIC/ACM international conference on web intelligence. pp 184–190

Viegas F, Wattenberg M, Kriss J, van Ham F (2007) Talk before you type: Coordination in Wikipedia. In: Proceedings of the 40th Hawaii international conference on system sciences

Voss J (2005) Measuring Wikipedia. In: Proceedings of the international conference international society for scientometrics and informetrics (ISSI '05), Stockholm

Wilkinson D, Huberman B (2007) Assessing the value of cooperation in Wikipedia. First Monday 12(4)

Wöhner T, Peters R (2009) Assessing the quality of Wikipedia articles with lifecycle based metrics. In: WikiSym '09: Proceedings of the 5th international symposium on wikis and open collaboration. ACM, New York, pp 1–10

Zheleva E, Sharara H, Getoor L (2009) Co-evolution of social and affiliation networks. In: KDD '09: Proceedings of the 15th ACM SIGKDD international conference on Knowledge discovery and data mining. ACM, New York, pp 1007–1016

Zlatic V, Bozicevic M, Stefancic H, Domazet M (2006) Wikipedias: Collaborative web-based encyclopedias as complex networks. Phys Rev E 74(1):016115

Chapter 5
Bridging the Gap Between Computational Models and Viability Based Resilience in Savanna Ecosystems

Justin M. Calabrese, Guillaume Deffuant, and Volker Grimm

5.1 Introduction

Savanna ecosystems cover a substantial percentage of Earth's land surface area and are economically and socially important (Scholes and Archer 1997; Sankaran et al. 2005; Bond 2008). In addition to harbouring considerable biodiversity, savannas are used extensively by humans, primarily as grazing lands for cattle (Scholes and Archer 1997; Vetter 2005). A growing percentage of the world's population depends on savannas for their livelihood, and as human populations grow, increasing pressure on savannas could lead to degradation (Vetter 2005).

Grazing can have a strong effect on the condition of the savanna. As fires are fuelled by grass, grass biomass affects both the frequency and intensity of fires in the savanna (van Wilgen et al. 2000). Fire is thought to regulate tree and shrub populations primarily by inhibiting the establishment of fire-sensitive juveniles (Pellew 1983; Higgins et al. 2000). If the savanna is overgrazed, there will be insufficient grass biomass to fuel fires and woody vegetation can begin to take over–a condition known as bush encroachment (see also Fig. 2.1 in Chap. 1).

J.M. Calabrese (✉)
Smithsonian Conservation Biology Institute, National Zoological Park, 1500 Remount Rd., Front Royal, VA 22630 USA

Helmholtz Centre for Environmental Research-UFZ, Permoserstrasse 15, 04318 Leipzig, Germany
e-mail: CalabreseJ@si.edu

G. Deffuant
Cemagref - LISC, 24 av. des Landais 63172 Aubière, France
e-mail: guillaume.deffuant@cemagref.fr

V. Grimm
Helmholtz Centre for Environmental Research-UFZ, Permoserstrasse 15, 04318, Leipzig, Germany
e-mail: volker.grimm@ufz.de

G. Deffuant and N. Gilbert (eds.), *Viability and Resilience of Complex Systems*, Understanding Complex Systems, DOI 10.1007/978-3-642-20423-4_5,

Once woody vegetation dominates, recovering the savanna's grazing value may be effectively impossible on time scales that matter to humans, given that savanna trees routinely live for greater than 100 years (Vetter 2005).

The widespread and growing threat of bush encroachment has been studied from both theoretical and applied perspectives, and various management approaches have been developed to cope with it. A clear consensus on the effectiveness of alternative strategies has, however, yet to emerge (Vetter 2005). To date, most management recommendations for savannas have been based on rangeland models, which are minimalistic theoretical models typically aiming to isolate a small number of potential driving mechanisms in savanna dynamics and bush encroachment (Perrings and Walker 1997; Anderies et al. 2002; Campbell et al. 2006). Their simplicity facilitates thorough analysis (e.g., Anderies et al. 2002), but often at the cost of being so "poor" in structure and mechanism (DeAngelis and Mooij 2003) that a thorough validation against multiple observed patterns is impossible (Grimm et al. 2005).

Though rangeland models are useful tools, they may oversimplify savanna dynamics. In particular, as they tend to be spatially-implicit models, they do not account for important spatial processes such as dispersal, competition, and the spread of fire. A potential alternative to rangeland models are the "computational" savanna models. These models attempt to capture the essential elements and processes in savanna ecosystems while still being relatively parsimonious (Pellew 1983; Jeltsch et al. 1996; Higgins et al. 2000). They are typically spatially-explicit and are often used to coherently integrate available empirical information at particular focal sites (e.g. Menaut et al. 1990; Jeltsch et al. 1996). Furthermore, computational savanna models often agree well with empirical patterns such as tree-grass ratios and observed tree spatial distributions (e.g., Jeltsch et al. 1998, 1999; Higgins et al. 2000). Table 5.1 gives a partial listing of computational savanna models and their focal sites. The main drawback of these models is that their complexity makes them challenging to parameterize and analyze.

The dynamic management of complex systems such as savannas is a difficult endeavour for at least two reasons. First, many existing schemes for defining management strategies focus on the system's attractors, and thus emphasize system behaviour near equilibria (Martin 2004). This steady-state focus stands in stark contrast to the empirical reality that complex systems needing management are typically disturbed and are operating far from their equilibria (Tietjen and Jeltsch 2007). Second, dynamic models that are detailed enough to reproduce the important aspects of the system's behaviour are not trivial to develop and are typically difficult to analyze. Such difficulties have precluded the use of computational savanna models as management tools.

Viability theory could potentially aid savanna management by addressing the first limitation mentioned above (Martin 2004). In the viability approach, available control actions are coupled to system dynamics. The ability of this coupled system to respect defined constraints on state variables of interest can then be studied and a control policy that best achieves these goals formulated. Viability theory also allows

Table 5.1 A partial list of computational savanna models available in the literature and the sites for which they were designed. If the model was not assigned a name by its authors, the last name of the first author of the relevant publication(s) is used

Name	Site	References
Baxter	Krueger National Park, South Africa	Baxter and Getz 2005
FLAMES	Kakadu National Park, Australia	Liedloff and Cook 2007
FORSAT	Kandara, East Cameroon	Favier et al. 2004
Higgins	Southern Africa	Higgins et al. 2000
Jeltsch	Kalahari Gemsbok National Park, South Africa	Jeltsch et al. 1996, 1997a Jeltsch et al. 1997b, 1998
Menaut	Lamto research station, Ivory Coast	Menaut et al. 1990
Pellew	Serengeti National Park, Tanzania	Pellew 1983
SATCHMO	Pniel Estates, South Africa	Meyer et al. 2007a Meyer et al. 2007b
SAVANNA DYNAMICS	Gr. Serengeti ecosystem, Tanzania and Kenya	Holdo et al. 2009
van Wijk	La Copita research area Texas, USA	van Wijk and Rodriguez-Iturbe 2002

a redefinition of the concept of resilience (see Chap. 1 for classical definitions) that is useful for highly disturbed systems operating far from their equilibria. This new definition emphasizes how readily the system can be managed given its current state and the allowable control actions (Martin 2004, Chap. 2). Chapter 2 demonstrates the analysis of a rangeland model, that of Anderies et al. (2002), in the viability framework.

Unfortunately, computational savanna models cannot be directly analyzed in the viability framework because they are too complex for currently available methods for calculating control policies and approximating viability kernels (see the dimensionality curse in Chap. 7). Thus, there exists a rift between computational savanna models and the tools that would potentially allow them to be used for management purposes.

We demonstrate here that this gap can be bridged by building a simplified approximating model that captures the key pattern dynamics of the full computational model. The approximating model can then serve as a proxy for the full model in the viability calculations. Our approach combines statistical inverse modelling with moment approximation techniques (sensu Matsuda et al. 1992), and yields a low-dimensional set of ordinary differential equations that captures the key properties of the full computational model. Coupling this simplified representation to newly developed tools for solving viability problems and computing action policies (Deffuant et al. 2007) allows us to close the loop between complex, system-specific computational models and viability theory. Throughout, we apply our approach to a particularly well developed example of the class of computational savanna models, that of Jeltsch and colleagues. (Jeltsch et al. 1996, 1997a, b, 1998).

5.2 The Jeltsch Model

5.2.1 Background and Overview

In the 1990s, Jeltsch and colleagues (Jeltsch et al. 1996, 1997a, b, 1998) built and analyzed a detailed savanna model based on a wealth of data from the Kalahari Gemsbok National Park in South Africa. The full details of the model and its construction can be found in the original references. A description of the model using Grimm et al.'s (2006) ODD protocol for describing computational models can be found on the PATRES project website (www.patres-project.eu) and a Windows executable version of the model is available from the authors upon request. Here, we give a brief overview of the key aspects of the model.

The Jeltsch model divides space into 5×5 m lattice sites, which is about the canopy size of the tree *Acacia erioloba* on which the model focuses. The model area consists of a 100×200 lattice of such sites, for a total area of 50 ha. Time proceeds in discrete steps of 1 year. Each year, the system receives as input some amount of rainfall. Rainfall input can be fixed, may vary stochastically, or may be based on observed rainfall time series. In all that follows, we will focus on the random rainfall scenario. Rain input affects soil moisture, which is divided into upper and lower layers. Five classes of vegetation are recognized: (1) trees, (2) shrubs, (3) perennial grasses, (4) annual grasses, and (5) mixed vegetation. Additionally, lattice sites may be completely empty (bare ground).

Within each year, the different vegetation types compete for moisture and space. Each type of vegetation is characterized by its own set of demographic parameters, which affect its ability to compete with other types. These parameters include per year establishment and death rates, which for trees are age-class specific. All grass types disperse globally, shrubs disperse only to adjacent sites, and trees employ a mixture of short-range, exponentially decaying dispersal and global dispersal. All types of vegetation can affect the soil moisture of the site in which they grow, but only trees can reduce the soil moisture in adjacent sites. Additionally, trees have greater access to soil moisture in the lower soil horizon compared to the other vegetation classes. Table 5.2 presents the user-tunable parameters of the implementation of the Jeltsch model used here. Grazing and fire may also occur, and the rules governing these disturbances will be described later.

Table 5.2 Reference parameter set for the Jeltsch model

Parameter name	Value	Parameter name	Value
Basic vegetation model		**Rainfall**	
Number of dispersed seeds	50	Rainfall scenario	Rand.
Number of seed patches	20	Avg. topsoil moist. (scale of 1–4)	3
Number of seeds per patch	20	Avg. subsoil moist. (scale of 1–4)	3
Tr max. establishment rate	0.6	**Initialization**	
Pr max. establishment rate	0.5	Initial Tr prop. cover	0.01
An max. establishment rate	1.0	Initial Pr prop. cover	0.5
Tr topsoil moisture influence	0.3	Initial An prop. cover	0.2
Pr topsoil moisture influence	0.3	**Fire and grazing**	
An topsoil moisture influence	0.4	Tr fire sens. (prob. of burning)	0.2
Tr subsoil moisture influence	0.6	Crit. fire thresh. (kg/ha)	275
Pr subsoil moisture influence	0.2	Crit. fire thresh. with graz. (kg/ha)	225
Tr death rate (<2 years old)	1.0	Grazing pressure (ha/lsu)	150
Tr death rate (2–3 years old)	0.8		
Tr death rate (4–5 years old)	0.6		
Tr death rate (6–10 years old)	0.2		
Pr death rate	0.6		

Abbreviations are: *Tr* tree, *Pr* perennial grass, *An* annual grass. See the original references for further details on the definition of model parameters

5.2.2 *Pattern Dynamics of the Jeltsch Model*

To simplify the state space of the Jeltsch model, we focus on only two of the above-mentioned vegetation types: trees and grasses. This approach is consistent with the model's original focus, as trees are modelled in much greater detail than the other vegetation types (Jeltsch et al. 1996). As such, we concern ourselves here with the population dynamics and spatial pattern formation of trees in the Jeltsch model. We will demand that the approximating model be capable of reasonably reproducing these features of the Jeltsch model.

Specifically, we will focus on three state variables: (1) tree density, (2) "near' neighbourhood spatial pattern, and (3) 'far' neighbourhood spatial pattern. We measure tree density as $\rho[1]$, the proportion of lattice sites occupied by trees. We quantify spatial pattern using the normalized pair correlation statistic defined (Stoyan and Stoyan 1994; Wiegand and Moloney 2004)

$$g_x = \frac{\sum N_x}{\lambda \sum A_x} \tag{5.1}$$

where λ is the density of trees in the modelled area, x refers to the inner radius of an annulus of width dx, N is the number of trees in the annulus, A is the area of the annulus and the sums are over all trees in the modelled area. In the lattice case treated here, the 'annulus' is a square ring one cell in width ($dx = 1$). Thus

g_1 refers to the ring of 8 sites immediately adjacent to a focal cell (i.e., the Moore neighbourhood), g_2 to the ring of 16 sites surrounding the Moore neighbourhood, and so on. For convenience we will refer to the 1 and 2 cell distances as the 'near' and 'far' neighbourhoods, respectively, and we use the subscripts n and f to identify quantities specific to these neighbourhoods. For example, the g statistics in the near and far neighbourhoods are denoted g_n and g_f, respectively. The normalized g statistic equals one for a random spatial distribution, with values greater and less than one indicating clustered and regular distributions, respectively (Stoyan and Stoyan 1994; Wiegand and Moloney 2004).

Though quite complex, the Jeltsch model appears to behave as a typical Markov chain. For scenarios producing tree-grass coexistence, state variables of interest, including $\rho_{[1]}$, g_n and g_f, reach a stationary state where they fluctuate around a mean value after an initial transient phase. We therefore assume that the model is ergodic, and estimate the stationary values of focal variables by running the model for a large number of time steps, taking widely spaced samples (to break up correlations among samples) of these variables starting after the model has reached stationary state, and averaging these samples (Bolker et al. 2000). Transient values of state variables are estimated by averaging (at each point in time) multiple, independent realizations of the model with the same parameter values and initial conditions.

Our goal here is not to provide a complete and detailed analysis of the model (for that, see Jeltsch et al. 1996, 1997a, b, 1998), but rather to approximate the model at the parameter values deemed most relevant for the focal study site. We justify this focus by noting that in a real management problem, the site-specific model parameters would first need to be determined. All subsequent analyses would be contingent upon these estimated parameters. Table 5.2 lists the Jeltsch model parameter values on which we based our approximations. For a more complete description of model parameters, including those not accessible to the user, see Jeltsch et al. (1996) and (1997b).

5.2.2.1 Logistic-Like Tree Population Growth

The key result of examining time series produced by the Jeltsch model is that tree population dynamics can be approximated by logistic growth (Fig. 5.1). This is fortunate, given that the interactions influencing tree population growth in the full Jeltsch model are numerous and complex. The net result of all of these factors is that the mean behaviour of ensembles of runs from the Jeltsch model can be well described by the classical two-parameter (r, K) logistic growth equation (Fig. 5.1).

5.2.2.2 Characteristic Short-Distance Spatial Pattern

The spatial arrangement of trees in the Jeltsch model appears to be largely driven by the antagonism between very short distance competition and slightly longer distance dispersal. In particular, dispersal limitation promotes the clustering that

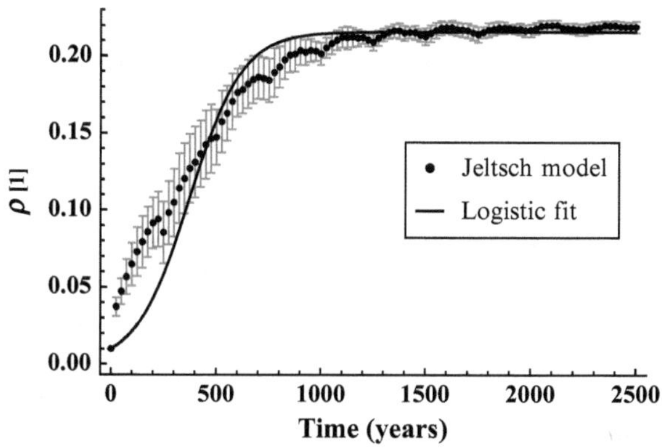

Fig. 5.1 Dynamics of $\rho_{[1]}$ in the Jeltsch model for the parameters listed in Table 5.2. The points are means and the error bars are standard deviations, both estimated from 30 independent realizations of the Jeltsch model. The classical logistic population growth model was fit to the Jeltsch model output by minimizing the sum of absolute deviations. The best fit parameter values were $r = 0.008$ and $K = 0.2156$

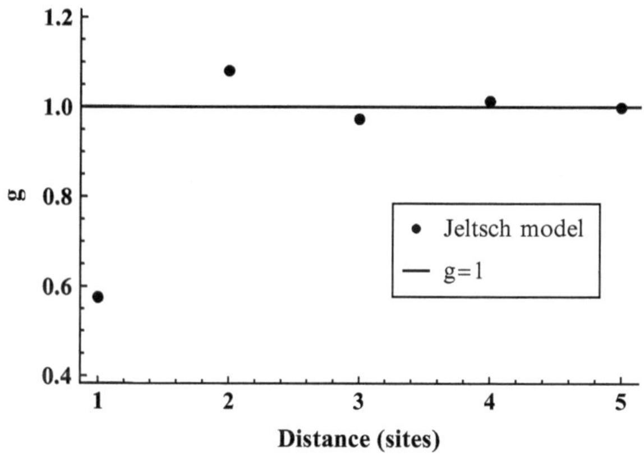

Fig. 5.2 Spatial pattern produced by the Jeltsch model for the parameters listed in Table 5.2. The combination of near neighbourhood regular spacing ($g_n < 1$), far neighbourhood aggregation ($g_f > 1$), and a rapid decay to randomness thereafter is a general feature of the Jeltsch model and does not depend strongly on user-tunable model parameters

is apparent at the far neighbourhood scale (Fig. 5.2, $g_f > 1$). Because trees can reduce the moisture levels of sites immediately adjacent to them (i.e., their near neighbourhood), trees tend not to occur adjacent to one another. This near neighbourhood repulsion effectively punches a hole in the clustering caused by dispersal, and leads to a strongly regular dispersion at the near neighbourhood scale (Fig. 5.2, $g_n < 1$). At larger spatial scales (g_{3+}), the patterns quickly decay to randomness, and therefore will not be discussed further. The features causing this two-scale spatial pattern are hard-coded into the model and were not varied across all of the analyses performed by Jeltsch and colleagues. They will, therefore, likely always be present regardless of the values of user-tunable parameters.

5.2.2.3 Summary of Pattern Analysis

These analyses suggest that the Jeltsch model could be usefully approximated by a simpler stochastic spatial model. Specifically, the combination of logistic-like population growth and two-scale spatial pattern suggests that a spatial (lattice) logistic model would be appropriate (e.g., Calabrese et al. 2010; Vazquez et al. 2010). The short-distance spatial interactions demonstrated in Fig. 5.2 suggests a spatial logistic model with two nested interaction scales. A model where both competition and dispersal act over the smaller scale, while only dispersal acts on the larger scale, should be able to capture the qualitative pattern demonstrated in Fig. 5.2.

The spatial logistic framework is ideal for our purposes here because equation-based approximations using moment-closure techniques have been developed for this class of models (Matsuda et al. 1992; Ellner 2001). Thus if we focus on an extension of the spatial logistic model that incorporates the above described features, it should be able to: (1) capture the key pattern dynamics of the Jeltsch model, and (2) be approximated using moment equation methods, yielding a low-dimensional system of ordinary differential equations. These approximations would then provide a link between Jeltsch model and existing methods for computing action policies from viability theory.

5.3 A Simplified Spatial Savanna Model

5.3.1 Description of the Simplified Model

The two-scale lattice logistic savanna model developed by Calabrese et al. (2010) incorporates the above-described features and thus will serve as a basis for our approximating model. A full description and detailed analysis of the model together with derivations of moment-equation-based approximations of the model can be found in Calabrese et al. (2010), while we give a brief overview here.

The approximating model is an extension of the contact process (Marro and Dickman 1999) and is implemented as a continuous time, discrete state Markov chain on a square lattice with periodic boundary conditions. Each lattice site is either occupied by an adult tree (state 1) or by grass (state 0). The proportion of tree-occupied sites on the lattice is denoted $\rho_{[1]}$, and the proportion of grass-occupied sites is $1 - \rho_{[1]}$. As in the Jeltsch model, we assume each site is 5×5 m and model a lattice of 100×200 sites, for a total area of 50 ha.

5.3.1.1 Interaction Neighbourhoods

The simplified model focuses on spatially limited seed dispersal, and on how competition among trees and fire affect the establishment chances of new trees. Dispersal and competition are spatially explicit processes and occur over defined neighbourhoods. As in the Jeltsch model, we use Moore neighbourhoods and define the near neighbourhood to be the $z_n = 8$ sites immediately surrounding the focal site and the far neighbourhood as the $z_f = 16$ sites surrounding the near neighbourhood. The dispersal and competition neighbourhoods will be built from these components.

5.3.1.2 Death

Adult trees (state 1 sites) die at a constant rate α, independent of their neighbourhood status.

5.3.1.3 Birth/Dispersal

Based on our analysis of the Jeltsch model, we assume that the spatial scale of dispersal is larger than that of competition, and thus the dispersal neighbourhood comprises both the near and far neighbourhoods. Each site within an adult tree's dispersal neighbourhood receives seeds at rate $\beta = b/(z_n + z_f)$ independent of its occupancy status, where b is the per tree birth rate. Seeds that land on tree-occupied sites die, while those landing on empty (grass-occupied) sites have a chance to establish.

5.3.1.4 Establishment

A seed landing in an empty site establishes with probability $P_E = P_C^{Surv} P_F^{Surv}$, where P_C^{Surv} is the probability of surviving competition with adult trees and P_F^{Surv} is the probability of surviving fire. For now, we let the probability of surviving fire serve as a place holder by setting $P_F^{Surv} = 1$. We will specify a functional form for this component of the model later, after we consider how fire works in the Jeltsch model. The probability that an establishing tree survives competition with its adult neighbours is $P_C^{Surv} = e^{-\delta S}$, where S is the number of adults in its near neighbourhood and δ scales the strength of competition.

5.3.2 Analytical Approximations of the Simplified Model

Calabrese et al. (2010; see their Appendix B for a detailed derivation) describe a multi-scale pair approximation (MSPA; Ellner 2001) to the above-defined stochastic model that accounts for both the mean tree density and the spatial dependencies of the mean. The MSPA is written in terms of three different kinds of site occupancy frequencies. The singlet frequency, $\rho_{[i]}$, is the global unconditional probability of a site being in state i. The pair frequency, $\rho_{[ij]}$, is the probability that a randomly selected site is in state i and a randomly selected neighbour is in state j. The conditional pair frequency, $q_{[j/i]}$, is the probability that a neighbouring site is in state j given that the focal site is in state i. The MSPA can then be written as the coupled, closed system

$$\begin{aligned} \frac{d\rho_{[1]}}{dt} &= \beta\left(z_n q_{n[1/0]} + z_f q_{f[1/0]}\right)(1-\rho_{[1]}) \\ &\quad \times P_F^{Surv} e^{-\delta z_n q_{n[1/0]}} - \alpha\rho_{[1]} \\ \frac{1}{2}\frac{d\rho_{n[11]}}{dt} &= \beta\left(1 + (z_n - 1)q_{n[1/0]} + z_f q_{f[1/0]}\right)(\rho_{[1]} - \rho_{n[11]}) \\ &\quad \times P_F^{Surv} e^{-\delta(1+(z_n-1)q_{n[1/0]})} - \alpha\rho_{n[11]} \\ \frac{1}{2}\frac{d\rho_{f[11]}}{dt} &= \beta\left(z_n q_{n[1/0]} + 1 + (z_f - 1)q_{f[1/0]}\right)(\rho_{[1]} - \rho_{f[11]}) \\ &\quad \times P_F^{Surv} e^{-\delta z_n q_{n[1/0]}} - \alpha\rho_{f[11]}, \end{aligned} \tag{5.2}$$

where the near and far neighbourhood conditional site frequencies are given by

$$q_{n[1/0]} = \frac{\rho_{[1]} - \rho_{n[11]}}{1-\rho_{[1]}} \quad \text{and} \quad q_{f[1/0]} = \frac{\rho_{[1]} - \rho_{f[11]}}{1-\rho_{[1]}}. \tag{5.3}$$

The normalized pair correlation statistics (5.1) for the near and far neighbourhoods can be obtained from the MSPA by dividing the appropriate pair frequency by the squared tree density:

$$g_n = \frac{\rho_{n[11]}}{\rho^2_{[1]}} \quad \text{and} \quad g_f = \frac{\rho_{f[11]}}{\rho^2_{[1]}}. \tag{5.4}$$

As there are no spatial processes in the model acting beyond the far neighbourhood, the g statistic for neighbourhoods three sites away and farther, g_{3+}, in the MSPA equals one (Calabrese et al. 2010). Thus with g_n and g_f we have a direct link between variation in model parameters and scale-dependent tree spatial pattern.

Finally, a mean-field approximation, covering only the dynamics of $\rho_{[1]}$, is obtained by replacing the conditional local densities defined in (5.3) with the global, unconditional tree density, $\rho_{[1]}$, yielding a closed equation in $\rho_{[1]}$:

$$\frac{d\rho_{[1]}}{dt} = bP_F^{Surv} e^{-\delta z_n \rho_{[1]}}(\rho_{[1]} - \rho^2_{[1]}) - \alpha\rho_{[1]}. \tag{5.5}$$

5.4 Fitting the MSPA to the Jeltsch Model

5.4.1 Estimating Parameters Without Fire and Grazing

The simplified model has three free parameters (b, δ, and α) whose values must be determined. To do this, we first generated 30 independent, 2500 year time series from the Jeltsch model at the parameter values listed in Table 5.2. From the raw output of the Jeltsch model, we then calculated time-series of $\rho_{[1]}$, g_n, and g_f for each run. We estimated b, δ, and α in the MSPA by minimizing the sum of absolute deviations between $\rho_{[1]}$, g_n, and g_f in the simplified model and those calculated from the output of the Jeltsch model. Figure 5.3 shows the means and standard deviations across the 30 runs of the Jeltsch model, together with the corresponding best fit of the MSPA. Notice that the MSPA is not fully capable of reproducing the

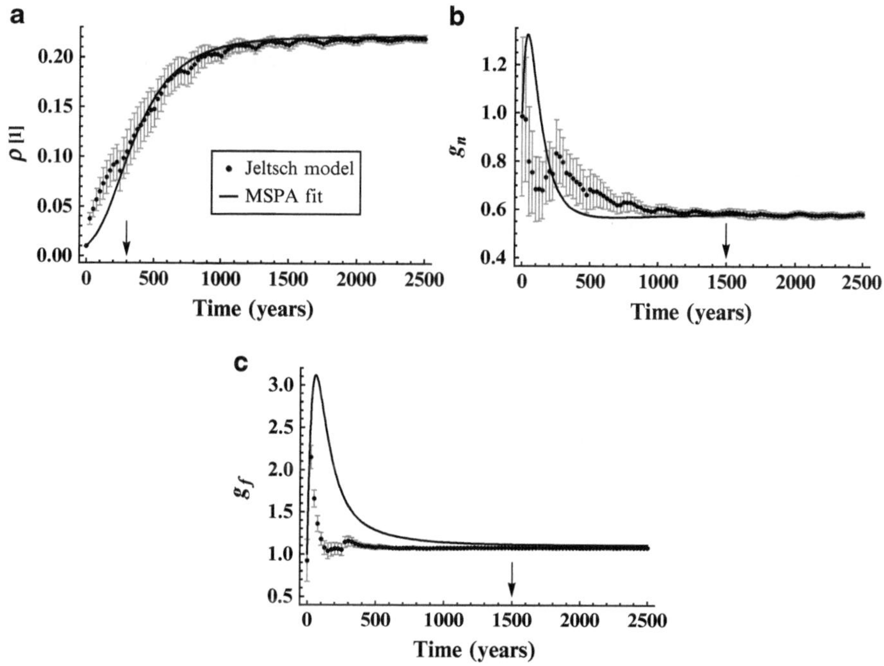

Fig. 5.3 The fit of the MSPA to the Jeltsch model for $\rho_{[1]}$ (Panel **a**), g_n (Panel **b**), and g_f (Panel **c**). The Jeltsch model parameters are given in Table 5.2, while the fitted values of MSPA parameters are given in Table 5.3. The transient dynamics $\rho_{[1]}$, g_n, and g_f in the Jeltsch model are qualitatively different than those of the MSPA, particularly for the spatial pattern indices. To avoid biasing the fit by forcing the MSPA to fit each time-series in its entirety, we instead only used the portions of each time series the MSPA was capable of fitting, which are those to the right of the *vertical arrows* in each panel. This resulted in good fits to the stationary values of each of the three variables, and also partially recovers the transient dynamics of $\rho_{[1]}$

Table 5.3 Parameter values for the approximation of the Jeltsch model by the MSPA. "Estimated" means that the parameter was estimated by minimizing the sum of absolute deviations between the MSPA and the output of the Jeltsch model. "Calculated" means that the parameter was directly calculated from the other model parameters

Parameter	Definition	How obtained?	Value
b	Tr growth rate	Estimated	0.014
δ	Tr competition intensity	Estimated	1.0428
α	Tr death rate	Estimated	0.0014
w	Fire survival shape	Estimated	18.6663
h	Grazing pressure	Estimated	0.0328
a	Scaling factor	Calculated	917.508

transient dynamics observed in the Jeltsch model. This is likely because when a new run is initialized in the Jeltsch model, all of the trees are young and randomly spaced, and so suffer very low mortality rates until they are > 100 years old. The influence of this initial, artificially low mortality phase dies out after a few hundred years. To account for this qualitative difference and avoid unnecessary bias in the parameter estimates, we fit the MSPA to a subset of the full times series for each variable, the beginning of which is denoted by an arrow in Fig. 5.3. As Fig. 5.3 demonstrates, the fitted MSPA can closely approximate the stationary values of all three variables, and recovers much of the transient dynamics of $\rho_{[1]}$ as well. Table 5.3 lists the estimates of b, δ, and α.

It is worth noting that due to the structure of the model, the values of the separate state variables cannot be tuned independently. In other words, changing the value of either the birth rate, death rate, or competition parameter affects the numerical values of all three state variables.

5.4.2 *Adding Fire and Grazing*

All of the above applies to the basic vegetation model but ignores fire and grazing. These two additional features represent key tools available to land managers to manipulate savannas, and they need to be in the model so that we can define sensible control actions on which the viability analysis can operate. The occurrence of fire in the Jeltsch model is determined primarily by a fire threshold parameter, ϕ. This parameter sets the level of grass biomass, measured in kilograms per hectare, above which fires may potentially occur. Above this threshold, the probability of fire occurrence is a saturating function of grass biomass (Jeltsch et al. 1996).

The filled circles in Fig. 5.4b demonstrate the effect of manipulating the fire threshold parameter on stationary tree density in the Jeltsch model with all other parameters held constant. Stationary tree density decreases slightly as the fire threshold is decreased (which results in increased fire frequency) until a phase transition

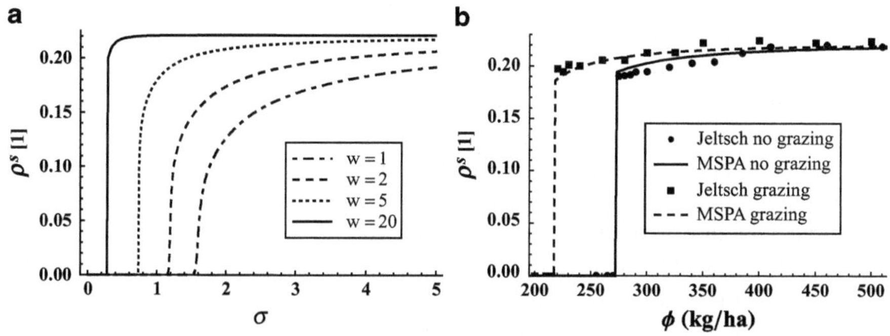

Fig. 5.4 The fire-induced phase transition in both the MSPA and the Jeltsch model. Panel **a** shows the differences in phase transitions in the MSPA caused by variation of the fire survival shape parameter. The *solid curve* in Panel **b** shows the fit of the MSPA (with (5.6)) to the phase transition observed in the Jeltsch model in the absence of grazing. The fit involves estimating one new parameter, w, from the Jeltsch model output. The *dashed curve* in Panel **b** considers the phase transition under the effects of grazing. Fitting the MSPA when grazing is involved requires the estimation of h, the grazing pressure. Estimated values of all MSPA parameters are in Table 5.3

occurs and the tree population collapses abruptly. This apparently discontinuous phase transition suggests an all or nothing effect of fire.

We tried various simple functional forms for the P_F^{Surv} term in the simplified model to approximate this behaviour. The one that provided the best compromise between simplicity and agreement with the behaviour of the Jeltsch model was

$$P_F^{Surv} = 1 - I\left[(1-h)(1-\rho_{[1]}); w, 1/\sigma\right] \tag{5.6}$$

where $I\,[z; x, y]$ is the regularized beta function, w controls the shape of the P_F^{Surv} curve and the location of the phase transition, and σ affects the propensity for the savanna to burn and is analogous to the fire threshold parameter in the Jeltsch model. Equation (5.6) is one minus the cumulative distribution function of the Beta distribution and is sometimes used in survival analysis (Lynch and Fagan 2009). The behaviour of the fitted MSPA with P_F^{Surv} given by (5.6) is demonstrated in Fig. 5.4a for various values of w.

Because the fire threshold in the Jeltsch model, ϕ, and the parameter σ in the approximating model are measured on different scales, it is necessary to define a mapping between them so that the phase transition functions produced by the two models can be directly compared. We aligned the two functions at the phase transition and then used a simple linear rescaling to transform σ to ϕ. The scaling factor is calculated as $a = \phi_c/\sigma_c$ where the c subscripts indicate the critical value at which the phase transition occurs in each model.

Having put both phase transitions functions on the same scale, we can estimate the value of the unknown shape parameter w. We do this by keeping the other MSPA parameters (b, δ, and α) fixed at their previously estimated values, and numerically searching for the value of w that minimizes the sum of absolute deviations between the phase transition function of the MSPA and that of the Jeltsch model. Table 5.3 lists the estimated value of w, and the solid curve in Fig. 5.4b shows the best fit of the MSPA to the phase transitions function observed in the Jeltsch model. Thus, capturing the behaviour of fire in the Jeltsch model requires the estimation of one additional parameter, w, bringing the total to four.

Grazing in the Jeltsch model is defined in terms of the stocking rate (ha/lsu) and affects the biomass of grass in the savanna. The amount of grass in the MSPA can be similarly modulated by tuning the parameter $0 \leq h < 1$. When $h = 0$, grazing is absent from the system and we obtain the solid curve in Fig. 5.4b. Adding grazing in either the MSPA or the Jeltsch model shifts the phase transition to the left, as the difference between the dashed/squares (grazing) and solid/circles (no grazing) sets of results in Fig. 5.4b demonstrate. Given the set of estimated MSPA parameters up to this point (b, δ, α, and w), we can estimate the value of h that minimizes the sum of absolute deviations between the phase transition function in Jeltsch model with grazing and that obtained from the MSPA. The dashed curve in Fig. 5.4b is the fitted MSPA with grazing, and the value of h used to obtain this fit is given in Table 5.3.

This completes the definition of the approximating model. In what follows, we focus on managing the savanna system defined by the approximating model using h, the grazing pressure, as a control.

5.5 Viability and Resilience Analyses

We shall devote the main part of our viability and resilience analysis to preventing tree density in the savanna from exceeding an arbitrarily chosen upper limit. This focus corresponds to managing a savanna to prevent bush encroachment, which is a problem typically faced by rangeland managers.

We first perform a viability and resilience analysis on the simpler but less accurate mean-field approximation of the model (5.5). After establishing baseline results in the mean-field case, we then move on to the pair approximation. We take variation of the grazing intensity, h, as the control action used to manage the system. We thus add the grazing intensity as a state variable, as was done in Chap. 2. The other model parameters are fixed according to the values of Table 5.3.

5.5.1 Mean Field Model

Before considering the possibility to manage the system by changing grazing pressure, it is worth studying the dynamics without action on grazing, in a similar approach as the one presented in Chap. 2. This study shows an attractor for a density $\rho_{[1]} = 0.22$, with a large attraction basin. If we rely on the usual mathematical definition of resilience and we suppose that this attractor value is desired, then we can define a desired set around $\rho_{[1]} = 0.22$, and we suppose that there is no

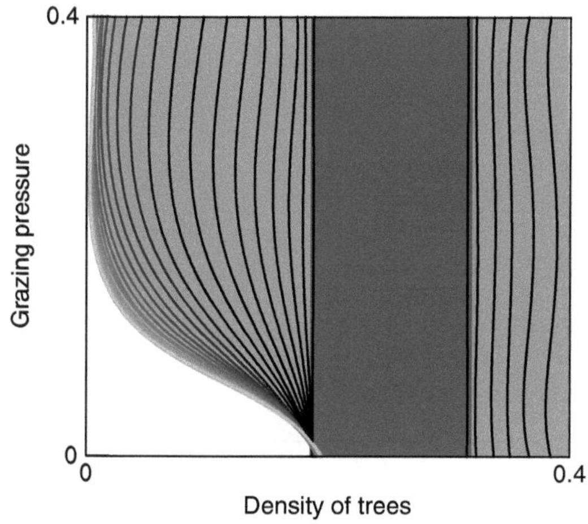

Fig. 5.5 Resilience basins when grazing does not change. The desired set is around the attractor value ($\rho_{[1]} = 0.22$). We note that the whole space is resilient, except a small area of low $\rho_{[1]}$ and low h (there is small attraction basin towards $\rho_{[1]} = 0$ in this area)

possibility of action (as we did in Chap. 2). To summarize, in this first experiment, the constraints are:

- The density of trees should be between two bounds: $0.18 \leq \rho_{[1]} \leq 0.32$.
- No constraint was imposed on the grazing, but for practical reasons (to limit the set of states to investigate) we chose: $0 \leq h \leq 0.6$. These limits have no influence on the viability.
- The grazing intensity never changes: $h' = 0$.

The result is given in Fig. 5.5. The darker zone corresponds to the desired set (which is equal to its viability kernel in this case), and the lines are the resilience basins, corresponding to increasing necessary time to go back to the desired area. We note that the system is quite resilient in this case, because the attraction basin of the attractor $\rho_{[1]} = 0.22$ covers almost the whole state space. The non-resilient states are in white at the bottom left of the figure and correspond to the attraction basin of the attractor $\rho_{[1]} = 0$ in this zone. Globally, it appears from this model that the attractor state is rather resilient.

Now, we can explore if it is possible, through appropriate management, to keep the system in a state other than its main attractor. Suppose that we want to keep the density of trees below the attractor value, and that we can modify the grazing pressure slightly (increasing or decreasing) at each time step. Moreover, we might want the grazing pressure to stay above a given threshold. In other words, we wish to maintain at least a minimal level of livestock production without causing bush encroachment. To summarize, we have now the following constraints defining the desired set, which yield the results presented in Fig. 5.6:

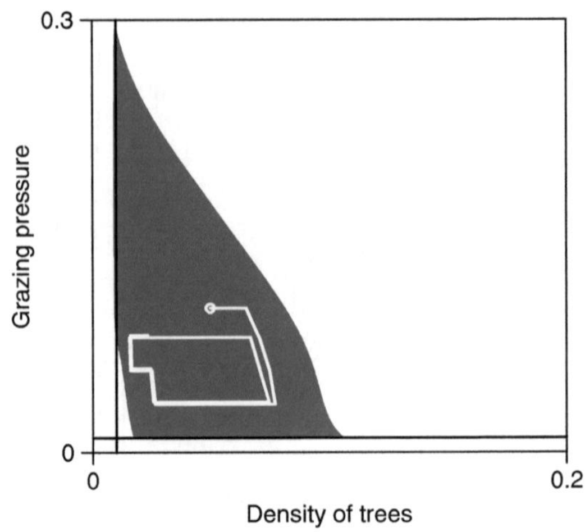

Fig. 5.6 Viability kernel for when the minimum acceptable level of grazing is 0.01 and the desired tree density is between 0.01 and 0.2. The viability kernel covers a part of the frontier between the attraction basins for attractors $\rho_{[1]} = 0.22$ and $\rho_{[1]} = 0$. The viable "lazy" trajectory makes cycles, crossing the limit between the attraction basins back and forth (because there is no attractor in the desired set)

- The density of trees should be between two bounds: $0.01 \leq \rho_{[1]} \leq 0.20$.
- We suppose that grazing pressure should satisfy: $0.01 \leq h \leq 0.3$.
- Grazing pressure can be changed at each time step: $|h'| < 0.0005$.

The results show that the corresponding viability kernel leads to management policies that make cyclic changes in grazing pressure. Grazing pressure must increase when tree density is too low, and must decrease when it is too high. The system then describes cycles within the viability kernel, as shown by the white trajectory in Fig. 5.6. This is a case of viable dynamics when there is no attractor in the desired set (another example is given in Chap. 2). This means that in the absence of management actions, the system would go to one of its attractors (depending on initial conditions) and would violate the viability constraints.

The resilience basins associated with this viability problem are shown in Fig. 5.7. The example trajectory shows that it is possible to go back to the viability kernel from situations where both the grazing pressure and the density of trees are too high. Doing so requires a decrease in the grazing pressure below the minimum acceptable level. Moreover, this is possible only in a limited part of the space, from which the attractor basin of $\rho_{[1]} = 0$ is reachable. This scenario implies that an overgrazed and partially bush encroached system could sometimes be recovered, but at the cost of severe and potentially long-term reductions in grazing pressure. Once the system ends up in the white area of Fig. 5.7, it has left the finite cost resilience basins and it is no longer possible to return to the desired set of states via the reduction of grazing intensity. Such a scenario corresponds to irreversible bush encroachment.

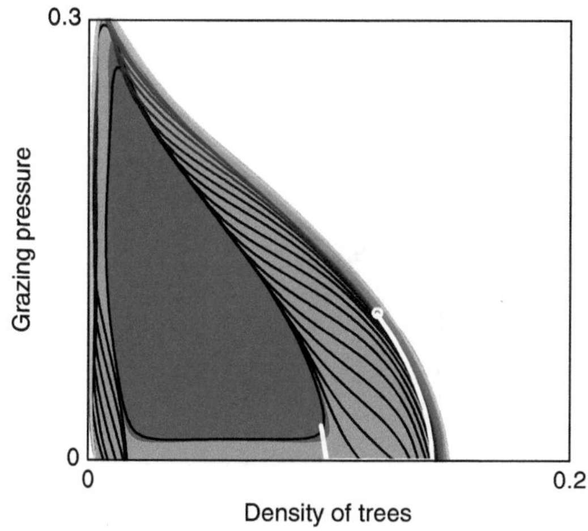

Fig. 5.7 Resilience basins corresponding to viability kernel of Fig. 5.6. The example of trajectory in light grey (starting at the *circle*) shows that to go back to the desired set, one must decrease the grazing as much as possible until getting below the required level of 0.01, and then wait until $\rho[1]$ moves below the viability kernel without modifying the grazing, and only then increase the grazing pressure to a level above 0.01 to drive the system back into the viability kernel

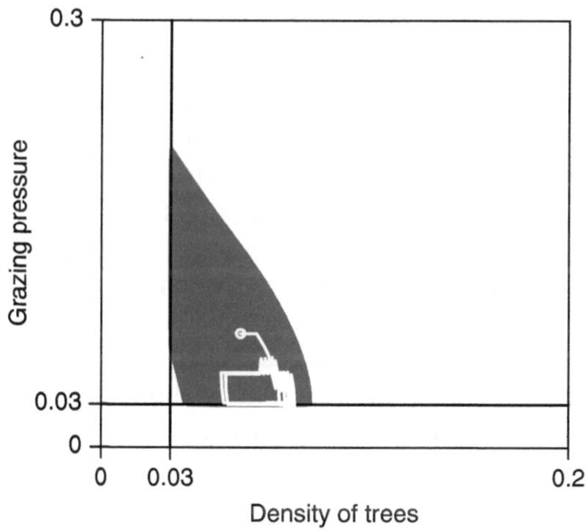

Fig. 5.8 Viability kernel for a minimum acceptable grazing pressure is higher than in Fig. 5.6 (0.03 vs. 0.01). As a result, the viability kernel is smaller and still covers the frontier between the attraction basins for $\rho[1] = 0.22$ and $\rho[1] = 0$. The example of trajectory also shows the necessity of cyclic management actions

If we increase the minimum acceptable level of grazing to 0.03, then we get the viability kernel presented in Fig. 5.8. We note that increasing the minimum level of grazing causes a significant decrease in the size of the viability kernel. The trajectory is also cyclic, but the changes in grazing pressure necessary to remain in the viability

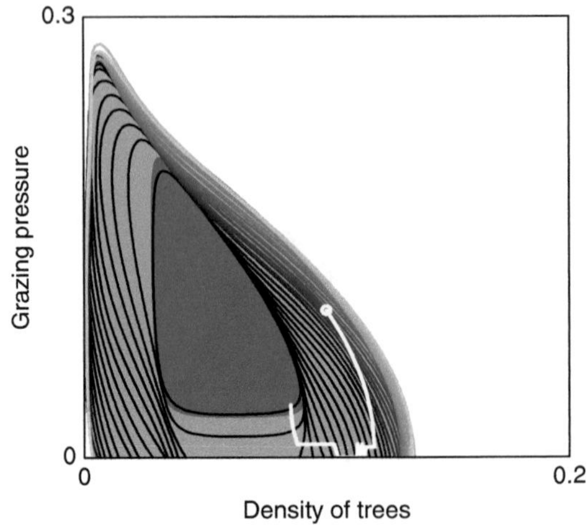

Fig. 5.9 Resilience basins corresponding to viability kernel of Fig. 5.8. The example of trajectory shows that to go back to the desired set, one must decrease the grazing, wait until the trajectory is below the viability kernel, and then increase the grazing pressure above 0.03

kernel must be much more frequent. Despite this, the set of resilient states is similar on the right side because the control policy required to get back to the viability kernel is the same.

The resilience basins associated with this viability kernel (see Fig. 5.9) are, however, very similar to resilience basins in Fig. 5.7, as are the trajectories that go back to the viability kernel. As before, the system is not resilient when it is no longer possible (at low grazing intensity) to reach the part of the space where the density of trees decreases. However, the resilience values are different, because the cost (time) to reach the viability kernel is higher when the minimum grazing is 0.03.

In this schematic problem, we have identified a classical tradeoff between the minimum level of grazing that should be guaranteed and the global sustainability of the system. When this minimum increases, the viability becomes more difficult to ensure, and requires more frequent management actions to be maintained.

5.5.2 *Pair Approximation Model*

The mean-field approximation is relatively crude, and it is interesting to investigate if running the viability approach on the MSPA leads to different conclusions. This is particularly interesting because the MSPA can account for some of the spatial correlations in tree density that are seen in real savannas and in many computational savanna models. To facilitate the viability analysis of the MSPA, it is convenient to have variables with a priori bounds. We therefore use slightly different variables than those previously described because the variables ρ_n [11] and ρ_f [11], have no such bounds. Hence, we prefer to use the new variables γ_n and γ_f, with:

$$\gamma_n = \frac{\rho_n[11]}{\rho[1]} \quad \text{and} \quad \gamma_f = \frac{\rho_f[11]}{\rho[1]}. \tag{5.7}$$

Indeed, we can always set $0 \le \gamma_n \le 1$ and $0 \le \gamma_f \le 1$. This defines completely the space within which to perform the calculations.

The dynamics on γ_n is given by:

$$\frac{d\gamma_n}{dt} = \frac{\frac{d\rho_n[11]}{dt}}{\rho[1]} - \frac{\frac{d\rho[1]}{dt}\rho_n[11]}{\rho^2[1]}. \tag{5.8}$$

This can be computed using (5.3), and the same can be done for γ_f. We performed the viability analysis under the same constraints as before:

- The desired density of trees is set between two bounds: $0.01 \le \rho[1] \le 0.2$.
- The grazing pressure is supposed to satisfy: $0.01 \le h \le 0.3$.
- The change of grazing at each time step is limited: $|h'| < 0.0005$.

Practically speaking, applying the software tool Kaviar on a problem of 4 dimensions such as this one (the dimensions are: $\rho[1]$, γ_n, γ_f and h), is more time consuming (this is the dimensionality curse, see Chap. 7). We used 19 points by dimension, which must be compared with the resolution of 61 points by dimension used in the 2D (mean-field) computations. However, 19 points per dimension leads to 130,321 points in 4D, while 61 per dimension results in only 3721 points in 2D. Moreover, the support vector machines defining the kernel or the resilience values typically require hundreds of support vectors in 4D (the one represented in Fig. 5.10 involves 468 support vectors), but only a few dozen in 2D.

Because the viability kernel is now 4-dimensional, we can only view 2D or 3D projections of it. When projected in the $(\rho[1], h)$ space for the different values of γ_n and γ_f that we tested, the viability kernel now appears substantially bigger than in the mean-field case (compare Figs. 5.6 and 5.8 with Figs. 5.10 and 5.11). Moreover, the example trajectory, demonstrating a "lazy" control strategy, shows that there is an attractor near $\rho[1] = 0.18, \gamma_n = 0.1, \gamma_f = 0.17, h = 0.022$. Indeed, after a while, the system oscillates around this point. It is worth noting that there is no equivalent attractor in the mean-field case. The much larger viability kernel and the presence of the new attractor suggest that the decision to use models that represent spatial correlations (e.g., the Jeltsch model and the MSPA) can fundamentally influence the management options that will keep the system in a desired range of states. In this particular case, analysis of the MSPA suggests that there are more management options available than the mean-field approximation would lead one to believe.

Figure 5.11 demonstrates the effects of raising both the minimum acceptable grazing pressure and minimum tree density ($0.03 \le \rho[1] \le 0.2$ and $0.03 \le h \le 0.3$). The viability kernel is much smaller than that obtained in Fig. 5.10, but it still contains the new attractor at $\rho[1] = 0.187$. In this case, however, the new attractor's basin of attraction is much smaller than before. These results imply that the set

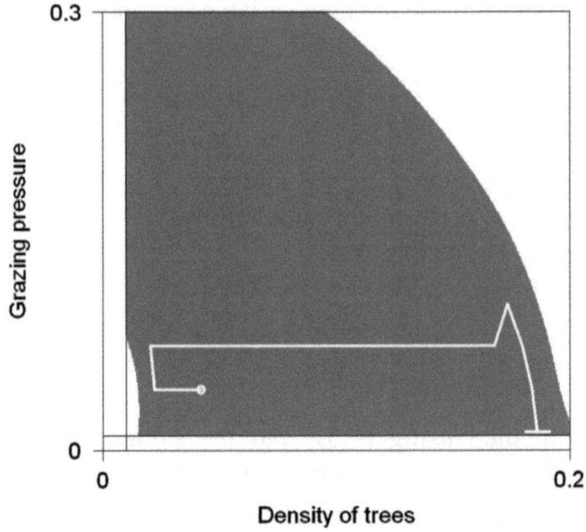

Fig. 5.10 A 2D projection of the 4D viability kernel for the MSPA, with constraints: $0.01 \leq \rho_{[1]} \leq 0.2$ and $0.01 \leq h \leq 0.3$, shown for $\gamma_n = 0.1$ and $\gamma_f = 0.1$. The trajectory begins at point: $\rho_{[1]} = 0.05, \gamma_n = 0.1, \gamma_f = 0.1, h = 0.05$ and finishes close to an attractor around: $\rho_{[1]} = 0.18, \gamma_n = 0.1, \gamma_f = 0.17, h = 0.022$

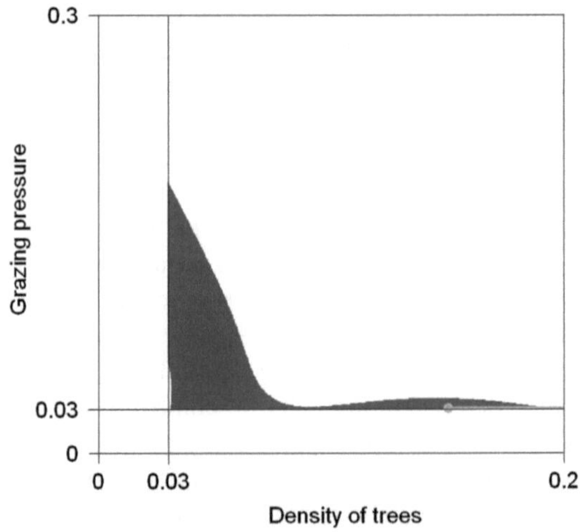

Fig. 5.11 A 2D projection of the 4D viability kernel for the MSPA, with constraints: $0.03 \leq \rho_{[1]} \leq 0.2$ and $0.03 \leq h \leq 0.3$, represented for $\gamma_n = 0.1$ and $\gamma_f = 0.17$. The trajectory begins at the point: $\rho_{[1]} = 0.15, \gamma_n = 0.1, \gamma_f = 0.17, h = 0.031$ and finishes close to an attractor around: $\rho_{[1]} = 0.187, \gamma_n = 0.11, \gamma_f = 0.18, h = 0.031$

of viable states can be strongly influenced by relatively subtle differences in the constraints imposed on the problem. Thorough analysis to understand how different sets of constraints affect the range of available management options is clearly warranted.

It appears that basing the viability analyses on the MSPA, which preserves some of the spatial correlations present in the full Jeltsch model, can lead to substantially different management options relative to the non-spatial mean-field approximation. However, the magnitude of these differences depends sensitively on the constraints, suggesting a need for careful choice of the constraints in applied management problems.

5.6 Discussion

Here, we have demonstrated that it is possible, at least in principle, to connect the dots between a model as complex as the Jeltsch model and viability theory. The current, rather severe, technical limitations of the algorithms used to approximate viability kernels and capture basins (see Chap. 7) demand that we find relatively simple approximations if we wish to analyze complex models in the viability framework. These approximations, which must capture key pattern dynamics of the original model if they are to be useful, can then be used as proxies in viability analyses.

We found that a simple lattice logistic model could approximate key aspects of tree population dynamics and spatial patterning in the considerably more complicated Jeltsch model. As models of this class are amenable to moment approximations (Matsuda et al. 1992; Ellner 2001), we were able to obtain equations describing the dynamics of our focal state variables. These equations, in turn, facilitated fitting the approximating model to the output of the Jeltsch model. Though the fitted approximating model agrees closely with the Jeltsch model, it would be preferable to mathematically define the mapping between the parameters of the two models. Unfortunately, the complexity of the Jeltsch model made such an approach intractable.

Switching fire on in the Jeltsch model does not change the values of the parameters governing the basic behaviours of different vegetation classes. We therefore chose to estimate the parameters relating to simple vegetation dynamics in the approximating model (b, α, and δ) with fire switched off in both models. With these estimated vegetation parameters held fixed, we then switched fire on in the models and estimated the fire-related parameters. This combined approach allowed us to define a simple approximation that captured the key dynamics of the Jeltsch model both with and without fire. The grazing intensity, h, which can affect tree recruitment indirectly by influencing fire frequency, then provides a reasonable control variable on which viability problems can operate.

Our viability analyses revealed sometimes striking differences between the mean-field and MSPA representations of the model. Viability kernels were in general larger when the MSPA was used, suggesting that the spatial correlations the MSPA preserves can have a pronounced effect on the range of viable management options. Furthermore, the viability analyses revealed the presence of an attractor inside the viability kernel that was completely absent in the mean-field case. These differences suggest that preserving spatial information in the models on which management decisions are based might be a key step in identifying the full range of options available to managers. As most current rangeland models are spatially-implicit (e.g., Perrings and Walker 1997; Anderies et al. 2002), further studies of how spatial processes influence the management picture that emerges from such models are warranted.

The computed viability results based on these simplified models are best viewed as a proof of concept at this point. Though it is possible to connect the Jeltsch model to viability theory, as we have demonstrated here, the path in-between is rather indirect. Using viability theory to define a sequence of management actions based on the Jeltsch model will be a much harder problem to solve than the one-way connection we have established here. There are two main reasons for this additional difficulty. The first is that to solve a real viability problem defined on the Jeltsch model, we would have to map the output of the Jeltsch model to the viability calculations, and the resulting control action back to the Jeltsch model in every time step of the viability calculations. Doing so would likely present considerable computational difficulties. The second issue is that the stochasticity that is present in the Jeltsch model but absent in the approximating models would likely affect the results. It is possible in principle to compute viability-based action policies for stochastic models, but doing so in practice would require modifying current methods and solving some technical problems.

Still, viability theory is relatively young and it is likely that the coming years will see further advances in the techniques used to estimate viability kernels and capture basins. In the meantime, our results suggest that it is possible to link quite complicated models to viability theory, albeit indirectly. We started with a model that was not designed for the purpose of being approximated by moment equations or being used in viability calculations. The model was, in this sense, not a "toy" model. When a viability analysis is a goal known at model design time, our results suggest that it would be possible to build models of intermediate complexity (relative to the Jeltsch model) that would be "mechanistically rich" enough (*sensu* DeAngelis and Mooij 2003) to capture the key behaviours of the real system, but would also lend themselves more readily to formal approximation techniques. Such models could be more tightly interfaced with viability calculations and might be able to serve as a basis for making sound management recommendations in real systems.

References

Anderies JM, Janssen MA, Walker BH (2002) Grazing management, resilience, and the dynamics of a fire-driven rangeland system. Ecosystems 5(1):23–44

Baxter PWJ, Getz WM (2005) A model-framed evaluation of elephant effects on tree and fire dynamics in African savannas. Ecol Appl 15(4):1331–1341

Bolker BM, Pacala SW, Levin SA (2000) Moment methods for ecological processes in continuous space. In: The geometry of ecological interactions: Simplifying spatial complexity. Cambridge

University Press, Cambridge, UK, Cambridge studies in adaptive dynamics, pages 388–411. Cambridge University Press, Cambridge, UK
Bond WJ (2008) What limits trees in C4 grasslands and savannas? Ann Rev Ecol Evol Systemat 39:641–659
Calabrese JM, Vazquez F, López C, San Miguel M, Grimm V (2010) The independent and interactive effects of tree-tree establishment competition and fire on savanna structure and dynamics. Am Nat 175:E44–E65
Campbell BM, Gordon IJ, Luckert MK, Petheram L, Vetter S (2006) In search of optimal stocking regimes in semi-arid grazing lands: One size does not fit all. Ecol. Econ. 60(1):75–85
DeAngelis DL, Mooij WM (2003) In praise of mechanistically rich models. In: Models in ecosystem science. Princeton University Press, Princeton, NJ, pp 63–82
Deffuant G, Chapel L, Martin S (2007) Approximating viability kernels with support vector machines. IEEE Trans. Automat Contr 52(5):933–937
Ellner SP (2001) Pair approximation for lattice models with multiple interaction scales. J Theor Biol 210(4):435–447
Favier C, Chave J, Fabing A, Schwartz D, Dubois MA (2004) Modelling forest-savanna mosaic dynamics in man-influenced environments: Effects of fire, climate and soil heterogeneity. Ecol Model 171(1–2):85–102
Grimm V, Berger U, Bastiansen F, Eliassen S, Ginot V, Giske J, Goss-Custard J, Grand T, Heinz SK, Huse G, Huth A, Jepsen JU, Jørgensen C, Mooij WM, Müller B, Pe'er G, Piou C, Railsback SF, Robbins AM, Robbins MM, Rossmanith E, Rüger N, Strand E, Souissi S, Stillman RA, Vab R, Visser U, DeAngelis DL (2006) A standard protocol for describing individual-based and agent-based models. Ecol Model 198(1–2):115–126
Grimm V, Revilla E, Berger U, Jeltsch F, Mooij WM, Railsback SF, Thulke HH, Weiner J, Wiegand T, DeAngelis DL (2005) Pattern-oriented modeling of agent-based complex systems: Lessons from ecology. Science 310(5750):987–991
Higgins SI, Bond WJ, Trollope WSW (2000) Fire, resprouting and variability: A recipe for grass-tree coexistence in savanna. J Ecol 88(2):213–229
Holdo RM, Holt RD, Fryxell JM (2009) Grazers, browsers, and fire influence the extent and spatial pattern of tree cover in the Serengeti. Ecol Appl 19(1):95–109
Jeltsch F, Milton SJ, Dean WRJ, van Rooyen N (1996) Tree spacing and coexistence in semiarid savannas. J Ecol 84(4):583–595
Jeltsch F, Milton SJ, Dean WRJ, Van Rooyen N (1997a) Analysing shrub encroachment in the southern Kalahari: A grid-based modelling approach. J Appl Ecol 34(6):1497–1508
Jeltsch F, Milton SJ, Dean WRJ, van Rooyen N (1997b) Simulated pattern formation around artificial waterholes in the semi-arid Kalahari. J Veg Sci 8(2):177–188
Jeltsch F, Milton SJ, Dean WRJ, van Rooyen N, Moloney KA (1998) Modelling the impact of small-scale heterogeneities on tree-grass coexistence in semi-arid savannas. J Ecol 86(5): 780–793
Jeltsch F, Moloney K, Milton SJ (1999) Detecting process from snapshot pattern: Lessons from tree spacing in the southern Kalahari. Oikos 85(3):451–466
Liedloff AC, Cook GD (2007) Modelling the effects of rainfall variability and fire on tree populations in an Australian tropical savanna with the FLAMES simulation model. Ecol Model 201(3–4):269–282
Lynch HJ, Fagan WF (2009) Survivorship curves and their impact on the estimation of maximum population growth rates. Ecology 90(4):1116–1124
Marro J, Dickman R (1999) Nonequilibrium phase tansitions in lattice models. Cambridge Univ Press, Cambridge, UK
Martin S (2004) The cost of restoration as a way of defining resilience: A viability approach applied to a model of lake eutrophication. Ecol Soc 9(2):8
Matsuda H, Ogita N, Sasaki A, Sato K (1992) Statistical mechanics of population–the lattice Lotka-Volterra model. Prog Theor Phys 88(6):1035–1049

Menaut JC, Gignoux J, Prado C, Clobert J (1990) Tree community dynamics in a humid savanna of the Ivory Coast – modeling the effects of fire and competition with grass and neighbors. J Biogeogr 17(4–5):471–481

Meyer KM, Wiegand K, Ward D, Moustakas A (2007a) The rhythm of savanna patch dynamics. J Ecol 95:1306–1315

Meyer KM, Wiegand K, Ward D, Moustakas A (2007b) Satchmo: A spatial simulation model of growth, competition, and mortality in cycling savanna patches. Ecol Model 209:377–391

Pellew RAP (1983) The impacts of elephant, giraffe and fire upon the *Acacia tortilis* woodlands of the Serengeti. Afr J Ecol 21(1):41–74

Perrings C, Walker B (1997) Biodiversity, resilience and the control of ecological-economic systems: The case of fire-driven rangelands. Ecol Econ 22(1):73–83

Sankaran M, Hanan NP, Scholes RJ, Ratnam J, Augustine DJ, Cade BS, Gignoux J, Higgins SI, Le Roux X, Ludwig F, Ardo J, Banyikwa F, Bronn A, Bucini G, Caylor KK, Coughenour MB, Diouf A, Ekaya W, Feral CJ, February EC, Frost PGH, Hiernaux P, Hrabar H, Metzger KL, Prins HHT, Ringrose S, Sea W, Tews J, Worden J, Zambatis N (2005) Determinants of woody cover in African savannas. Nature 438(7069):846–849

Scholes RJ, Archer SR (1997) Tree-grass interactions in savannas. Ann Rev Ecol Systemat 28:517–544

Stoyan D, Stoyan H (1994) Fractals, random shapes and point fields: Methods of geometrical statistics. Wiley, New York

Tietjen B, Jeltsch F (2007) Semi-arid grazing systems and climate change: A survey of present modelling potential and future needs. J Appl Ecol 44(2):425–434

van Wijk MT, Rodriguez-Iturbe I (2002) Tree-grass competition in space and time: Insights from a simple cellular automata model based on ecohydrological dynamics. Water Resour Res 38(9):1179–1193

van Wilgen BW, Biggs HC, O'Regan SP, Mare N (2000) A fire history of the savanna ecosystems in the Kruger National Park, South Africa, between 1941 and 1996. S Afr J Sci 96(4):167–183

Vazquez F, López C, Calabrese JM, Muñoz MA (2010) Dynamical phase coexistence: A simple solution to the "savanna problem". J Theor Biol 264(2):360–366

Vetter S (2005) Rangelands at equilibrium and non-equilibrium: Recent developments in the debate. J Arid Environ 62(2):321–341

Wiegand T, Moloney KA (2004) Rings, circles, and null-models for point pattern analysis in ecology. Oikos 104(2):209–229

Chapter 6
Viability and Resilience of a Bacterial Biofilm Individual-Based Model

Nabil Mabrouk, Jean-Denis Mathias, and Guillaume Deffuant

6.1 Introduction

In many industrial applications, bacteria are cultivated in mixed bioreactors filled with water and fed with nutrients. Depending on the reactor configuration and operating conditions, bacteria can either be grown as planktonic freely floating cells or attached to available surfaces forming a thin layer with a complex spatial structure called biofilm. In the planktonic growth mode, it is often assumed that all the bacterial cells experience the same average environmental conditions. Hence, their growth and metabolic activity can be controlled by acting directly on the environmental conditions in the reactor. Biofilm reactors are more difficult to control. Depending on the biofilm spatial structure, the local environment experienced by the individual bacterial cells in the biofilm may significantly differ from the conditions in the bulk phase of the reactor. It is therefore often difficult to predict how a change in the bulk phase conditions will impact the conditions inside the biofilm.

Many experimental observations of biofilm systems suggested that a biofilm can better be controlled by acting directly on the biofilm spatial pattern. The idea relies on the fact that bacteria forming the biofilm react to their local environment which to a large extent depends on the biofilm spatial structure. Hence a strategy to control the metabolic activity of the bacterial cells in a biofilm is to modify their local environment by disturbing the biofilm spatial structure.

In this chapter we propose to apply viability and resilience approaches to the control of biofilm spatial structure. The biofilm system is modelled using an individual-based model (IBM). IBMs are a powerful tool for simulating biofilm structure development (Grimson 1994; Kreft 2001; Picioreanu 2004; Helleweger 2009).

N. Mabrouk · J.-D. Mathias · G. Deffuant (✉)
Cemagref - LISC, 24 av. des Landais 63172 Aubière, France
e-mail: nabil.mabrouk@cemagref.fr; jean-denis.mathias@cemagref.fr; guillaume.deffuant@cemagref.fr

G. Deffuant and N. Gilbert (eds.), *Viability and Resilience of Complex Systems*, Understanding Complex Systems, DOI 10.1007/978-3-642-20423-4_6,

They simulate the development of microbial biofilms by specifying the behavioural and interaction rules at the level of the discrete individuals. IBMs can be helpful in exploring the richness of spatial patterns and in investigating how biofilm pattern and local heterogeneities form and impacts the individual-level dynamics (birth, death, migration, etc.). IBMs however evolve in a large parameter-state space which make them difficult to use for viability and resilience calculation. The strategy that we adopt in this chapter is to simplify and approximate the dynamics of the individual-based model through the derivation of a moment approximation model that we use to address viability and resilience problems.

The first section of this chapter is dedicated to the derivation of the individual-based model of a system formed with motile bacteria excreting polymeric substances that reduce their motility yielding the formation of a diversity of spatial patterns. The second section is dedicated to the simplification of the individual-based model and the derivation of the moment approximation model. We represent the dynamics of the moment model in a reduced space formed with three variables. The viability and resilience problems are treated in the last section of this chapter.

6.2 Individual-Based Modelling of Microbial Spatial Patterns

6.2.1 Main Principle: Self Excreted Polymer Reducing Bacteria Motility

We are interested in exploring the spatial patterns that arise in a system formed with motile bacteria colonizing a planar surface and excreting a product (exopolymer) that reduces their motility. Systems of this kind have been observed during laboratory cultivation of *Pseudomonas aeruginosa*, a model bacterium commonly used in biofilm research (Costerton 1995). The wild type of *P. aeruginosa* is able to move on a submerged solid surface. This surface-associated motility can be reduced by the accumulation on the surface of self-produced exopolymeric substances. Whitchurch et al. (2002) showed that the produced exopolymeric substances (in particular the extracellular DNA) released by the cells can bind the cells together and to the surface. The interplay between surface associated motility and exopolymer excretion in addition to competition on nutrient and detachment (bacteria removal) is believed to have an important role in the emergence of specific structured patterns.

The importance of these interactions in the emergence of biofilm spatial patterns can be explored using individual-based models. Mabrouk et al. (2009) proposed an IBM of a system formed with motile bacteria growing and dividing by up-taking a diffusive nutrient. They supposed that the individuals transform a part of the up-taken nutrient into a slowly diffusive polymeric product released in the environment and assumed that the excreted polymer reduces the individuals surface-associated motility. The motility of the bacteria is described as a Brownian process with a diffusion factor D_f expressed as a decreasing function of the local polymer concentration p:

$$D_f = \frac{D_{f0}}{1 + \beta p}, \tag{6.1}$$

with D_{f0} the maximum diffusion factor and β a positive parameter. Depending on the values of β, this system can yield a variety of patterns ranging from uniform distribution of the cells to patterns with interconnected microcolonies (Mabrouk 2009) (see Fig. 6.1). The model described there is designed to simulate the early stages of pattern formation with a high initial level of nutrient and neglected the death and detachment processes. In this section we are interested in the long-run biofilm dynamics and the development of stationary patterns. Hence we extend this IBM by adding a detachment process which may lead to a stationary state (for appropriate parameters). Several microbial IBMs represent the individuals as hard spheres (with a variable diameter) shoving each other. Such level of details provides a fine-scale description of the spatial patterns but increases the model complexity. Thus in the IBM presented here we do not make any assumption about the shape of the individuals and represent them as identical point particles.

We provide a detailed description of the IBM using the ODD (Overview, Design concepts and Details) framework. The IBM simulates the time evolution of the spatial pattern of the bacteria and the polymer particles. We analyze these patterns qualitatively and quantitatively by measuring the first and second spatial moments. In the second part on this section we propose to simplify the IBM by assuming that bacteria motility and detachment processes are dependent on the bacteria density rather than on the polymer density. The strategy of simplifying the complexity of the IBMs can be very beneficial to understand which processes produce the observed pattern. We assess in the third part of this section whether this simplification still allows us to reproduce the same diversity of patterns.

6.2.2 *Overview of the Detailed IBM*

6.2.2.1 State Variables and Scales

We consider a two-dimensional space containing two types of individuals (bacteria and polymeric particles) both represented as point particles. Each individual is fully characterized by its type and coordinate within the domain. The vectors $u = (x_i), i = 1..N_b$ and $v = (x_j), j = 1..N_p$ formed with the locations of all the bacteria and the polymeric particles, respectively, at a time t define the spatial pattern at that time.

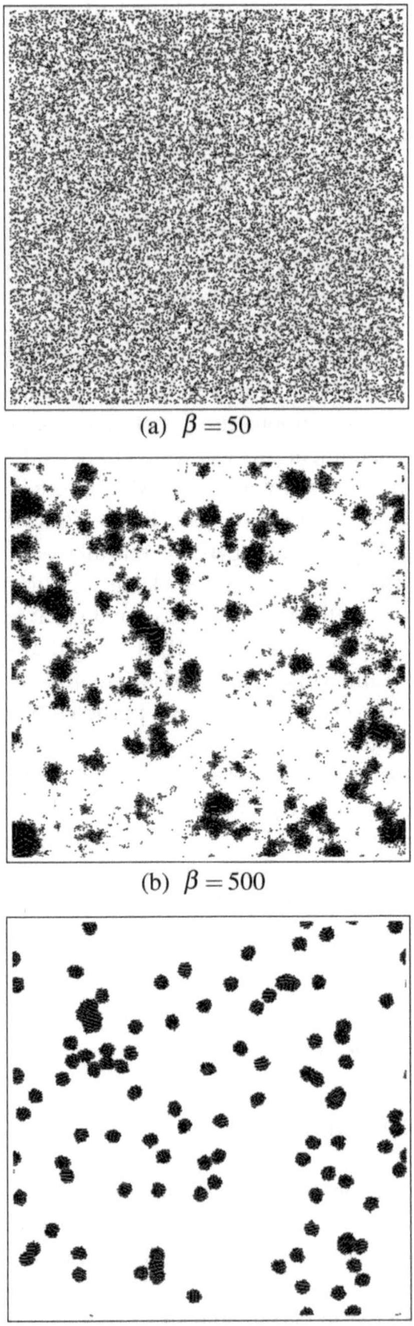

(a) $\beta = 50$

(b) $\beta = 500$

(c) $\beta = 5000$

Fig. 6.1 Snapshots of the transitory spatial pattern simulated with the IBM in Mabrouk (2009) for different values of β

6.2.2.2 Process Overview and Scheduling

We assume that the spatial pattern changes in time due to five stochastic events acting on the individuals:

- Birth of a new bacterium
- Detachment (removal) of a bacterium
- Migration of a bacterium to a new location
- Production of a new polymeric particle
- Detachment of a polymeric particle

Birth is a process by which a parent bacterium produces an offspring individual the location of which is selected randomly in the neighbourhood of the parent individual. Detachment is the process by which an individual is removed from the system. Polymeric particles are produced by the bacteria and can also be detached. The released polymeric particle is located in the neighbourhood of the mother bacterium. We suppose that bacteria detachment and motility as well as polymeric particles detachment are dependent on the local density of polymer such that a high polymer density reduces the migration of the bacteria and increases bacteria and polymer detachment probabilities.

The temporal behaviour of the IBM is governed solely by the five stochastic processes. We discretize the time with a constant time step Δt. At each time step we iterate over the following processes:

1. Calculate the detachment rates of the bacteria $D_i, i = 1..N_b$, the detachment rate of the polymeric particles D_j $j = 1..N_p$, and the motility rate of the bacteria $M_i, i = 1..N_b$
2. Perform bacteria birth events with a constant probability B_b
3. Perform the polymer production event with a constant probability B_p
4. Perform bacteria detachment events with a probability $D_i, i = 1..N_b$
5. Perform polymer detachment events with a probability $D_j, j = 1..N_p$
6. Perform bacteria motility events with a probability $M_i, i = 1..N_b$
7. Update the list of polymer and bacteria individuals
8. Update time according to $t = t + \Delta t$

6.2.3 *Details*

1. **Initialization:** The model is initialized with $N_b(t = 0) = 100$ bacterial cells distributed uniformly over the domain. We suppose that initially the domain contains no polymer.
2. **Submodels:**
   - Birth process: we suppose that the probability per unit of time that a bacterium particle i in position x_i produces a new bacterium located in position x' is given by:

$$B_b(x_i, x') = b_p K\left(\frac{||x_i - x'||}{w_{bb}}\right), \tag{6.2}$$

The parameter b_b is the birth rate. $K(||x_i - x'||/w_{bb})$ is called the birth kernel. This kernel gives the probability that the newly formed bacterium disperses instantaneously after the birth event to the location x'. For simplicity we use a uniform birth kernel, the general form of which is given by:

$$K\left(\frac{||x - x'||}{w}\right) = \begin{Bmatrix} 1/w & \text{if} \quad ||x - x'|| < w \\ 0 & \text{else} \end{Bmatrix} \tag{6.3}$$

- Polymer production process: we suppose that the probability per unit of time that a bacterium particle i in position x_i produces a new polymeric particle located in position x'' is given by:

$$B_p(x_i, x'') = b_p K\left(\frac{||x_i - x''||}{w_{bp}}\right), \tag{6.4}$$

The parameter b_p is the polymer production rate. $K(||x_i - x''||/w_{bp})$ is called the polymer production kernel. This kernel gives the probability that the newly formed polymeric particle disperses instantaneously after the production event to the location x''. We also use a uniform kernel (6.3) for the polymer production.

- Bacteria detachment process: the detachment probability per unit of time of a bacterium particle i located in x_i is given by D_i:

$$D_i = d_b + d'_b p_p(x_i, w_{db}), \tag{6.5}$$

d_b and d'_b are the density-independent and the density-dependent detachment rates. $p_p(x_i, w_{db})$ is the density of polymeric particles evaluated in x_i and given by:

$$p_p(x_i, w_{db}) = \sum_{j=1}^{N_p} K\left(\frac{||x_i - x_j||}{w_{db}}\right), \tag{6.6}$$

The local density of polymer in x_i is evaluated by summing the contribution of all polymer particles in the system, each weighted with an interaction kernel $K(||x_i - x_j||/w_{db}))$. We use a uniform bacteria detachment interaction kernel (6.3).

- Polymer detachment process: the detachment probability per unit of time of polymeric particle j located in x_j is given by D_j:

$$D_j = d_p + d'_p p_p(x_j, w_{dp}), \tag{6.7}$$

d_p and d'_p are the density-independent and the density-dependent detachment rates. $p_p(x, w_{dp})$ is the density of polymeric particles evaluated in x_j:

$$p_p(x_i, w_{dp}) = \sum_{k=1}^{N_p} K\left(\frac{||x_j - x_k||}{w_{dp}}\right), \tag{6.8}$$

Analogously to bacteria detachment we evaluate the influence of the neighbouring polymeric particles to the detachment of a focal polymeric particle using the uniform polymer interaction kernel $K(||x_j - x_k||/w_{dp}))$.

- Bacteria motility: we suppose that bacteria are motile and that the bacteria motility rate decreases linearly with the increase of the local polymer density. The probability of a bacterium i to move to a location x' is given by:

$$M(x_i, x') = M_i K\left(\frac{||x_i - x'||}{w_m}\right), \tag{6.9}$$

with M_i being the motion probability per unit of time of the bacterium in i, and $K(||x_i - x'||/w_m)$ is the motility kernel. The motion probability M_i is given by:

$$M_i = m_1 - m_2 p_p(x_i, w_v), \tag{6.10}$$

with m_1 and m_2 as the density-independent and the density-dependent motility rates and $p_p(x_i, w_v)$ the local polymer density in location x_i calculated using a uniform interaction kernel with side w_v.

3. **Parameters:** The model parameters are summarized in Table 6.1.
4. **Implementation:** We implement the model using the Mason framework, an open-source Java-based library for implementing agent-based models (Luke 2004). We conduct the numerical exploration of the IBM using the SimExplorer

Table 6.1 Model parameters

	Description	Dimension
b_b	Bacteria birth rate	$[T^{-1}]$
b_p	Polymer production rate	$[T^{-1}]$
d_b	Density-independent bacteria detachment rate	$[T^{-1}]$
d_p	Density-independent polymer detachment rate	$[T^{-1}]$
d'_b	Density-dependent bacteria detachment rate	$[T^{-1}]$
d'_p	Density-dependent polymer detachment rate	$[T^{-1}L^2]$
m_1	Density-independent bacteria motility rate	$[T^{-1}]$
m_2	Density-dependent bacteria motility rate	$[T^{-1}L^2]$
w_{bb}	Side of the bacteria birth kernel	$[L]$
w_{bp}	Side of the polymer production kernel	$[L]$
w_{db}	Side of the bacteria detachment kernel	$[L]$
w_{dp}	Side of the polymer detachment kernel	$[L]$
w_m	Side of the bacteria motility kernel	$[L]$
w_v	Side of the bacteria motility interaction kernel	$[L]$

software (www.simexplorer.org). SimExplorer is an open-source framework designed for managing simulation experiments. The simulation results are basically a list of the bacteria and the polymeric particle positions at each time step. We analyze these results qualitatively (by examining snapshots of the bacteria and polymer spatial pattern) and quantitatively by measuring two statistical quantities corresponding to the first and second spatial moments. The first spatial moments are given by the average densities of bacteria and polymer and contain no information about the spatial pattern. Second spatial moments are given by three pair correlation functions denoted C_{bb}, C_{bp} and C_{pp} measuring, respectively, the density of pairs of individuals formed with two bacteria, a bacterium and a polymeric particle, and two polymeric particles at different distances. Second spatial moments provide a quantitative description of the average neighbourhood of a focal individual. We normalize all pair correlation functions such that a value of one means that the individuals are distributed uniformly. Values higher than 1 indicate an aggregation process while values smaller than 1 indicate that the density of neighbours is lower than what would be expected if the individuals were uniformly distributed.

6.3 Spatial Patterns Generated by the Model

The IBM presented above has 8 kinetic parameters and 6 uniform kernels defining the process spatial scales. All these parameters can potentially impact on the spatial pattern. However, the numerical exploration of 14 parameters is computationally very expensive if not infeasible. Hence, we propose to fix the value of some parameters based on additional assumptions about the individual-level dynamics. We suppose that the bacteria birth and polymer production kernels are small compared to the size of the domain and take the same value for both kernels $w_{bb} = w_{bp}$. Furthermore, we suppose that polymer and bacteria detach at the same rate ($d_b = d_p$ and $d'_b = d'_p$) and we use the same interaction kernel for the calculation of the local polymer density $w_{db} = w_{dp}$ involved in the calculation of the detachment probabilities. With these assumptions, the number of parameters is reduced from 14 to 9 parameters as summarized in Table 6.2. Additionally, we fix the values of the birth rate, the birth kernel and the motility kernel. We focus on the impact of detachment and density-dependent bacteria motility on the observed pattern.

6.3.1 *Case 1: Immotile Bacteria ($m_1 = m_2 = 0$) with Density-Independent Detachment $b = d$ and $d' = 0$*

This is the simplest case where bacteria birth and detachment occur at the same constant rate independently of the polymer spatial distribution. Figure 6.2 shows the

Table 6.2 Model parameters

	Description	Value
L	Side of the squared spatial domain	201
Δt	Time step	1
b	Bacteria birth and polymer production rates	0.12
d	Density-independent bacteria and polymer detachment rate	Variable
d'	Density-dependent bacteria and polymer detachment rate	Variable
m_1	Density-independent bacteria motility rate	Variable
m_2	Density-dependent bacteria motility rate	Variable
w_b	Side of the bacteria birth kernel	3
w_d	Side of the bacteria detachment kernel	Variable
w_m	Side of the bacteria motility kernel	31
w_v	Side of the bacteria motility interaction kernel	Variable

evolution of a spatial pattern from uniformly distributed $N_b(t = 0) = 100$ bacterial cells to a pattern with colonies distributed randomly over the domain. The formation of these colonies is due to the small birth kernel. Newly formed cells are located in the neighbourhood of their parent cell. The size of the colonies increases with the increase of the size of the birth kernel (data not shown). For large birth kernels the colonies can not be distinguished and the bacteria are uniformly distributed.

The spatial pattern can be characterized by measuring the bacteria self pair correlation function. This function (denoted $C_{bb}(r)$) gives the average density of neighbouring bacteria at different distances from a focal one. Figure 6.3 shows the time evolution of the radial pair correlation function. Initially the function takes values close to 1 before peaking at short distances indicating the formation of colonies. The pair correlation function seems to reach a quasi-stationary state indicating that even though the colonies form and detach from the system, the individual experiences in average a quasi-stationary neighbourhood.

As bacteria birth and detachment are independent from the density of polymer ($d' = 0$) the polymer pattern has no impact on the bacteria dynamics. Polymeric particles are produced by the bacteria and accumulate (exclusively) within the colonies.

6.3.2 *Case 2: Immotile Bacteria ($m_1 = m_2 = 0$) with Density-Independent Detachment $b > d$ and $d' > 0$*

We investigate how density-dependent detachment affects the spatial pattern of the bacteria. The case $b = d$ and $d' > 0$ yields to the extinction of the bacteria population. Hence, we consider the case where the growth rate of the bacteria is higher than their density-independent detachment rate ($b > d$). Figure 6.4 shows the transitory and the (quasi) stationary spatial patterns. The bacteria are organized in colonies separated with a regular distance approximatively equal to the radii of the

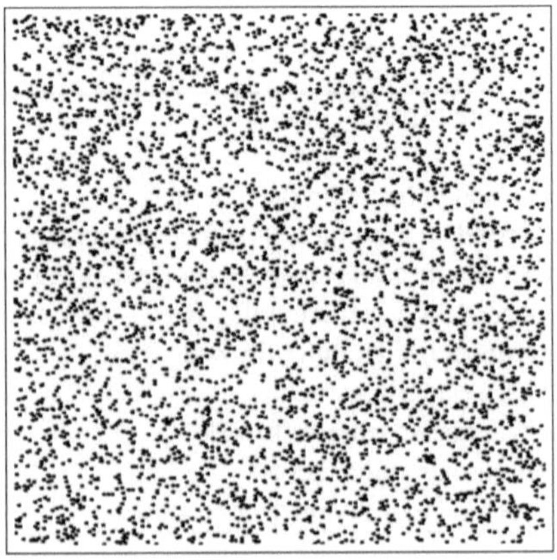

(a) $t = 0$

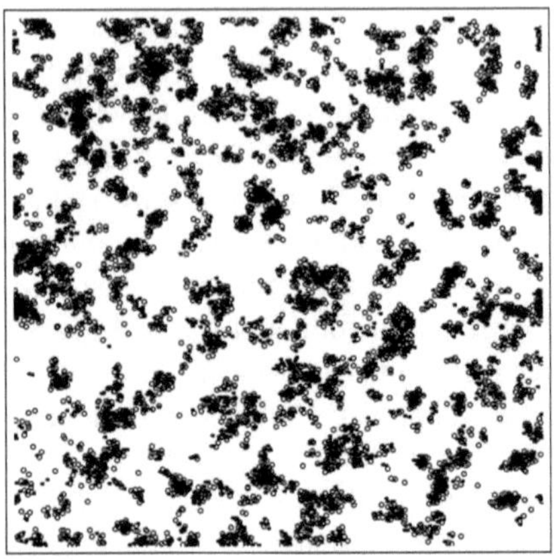

(b) $t = 200$

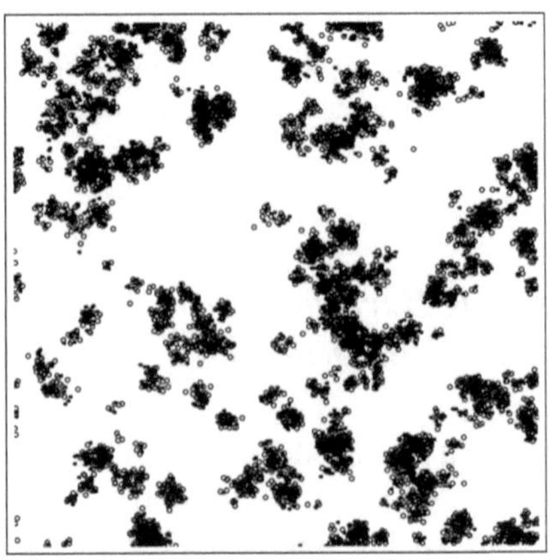

(c) $t = 1000$

Fig. 6.2 Snapshots of the time evolution of the spatial pattern. The simulation parameters are $b = 0.12$, $d = 0.05$, $d' = 0.0$, $m_1 = 0$, $m_2 = 0$, $w_b = 3$. *Black* bacteria, *White* polymer

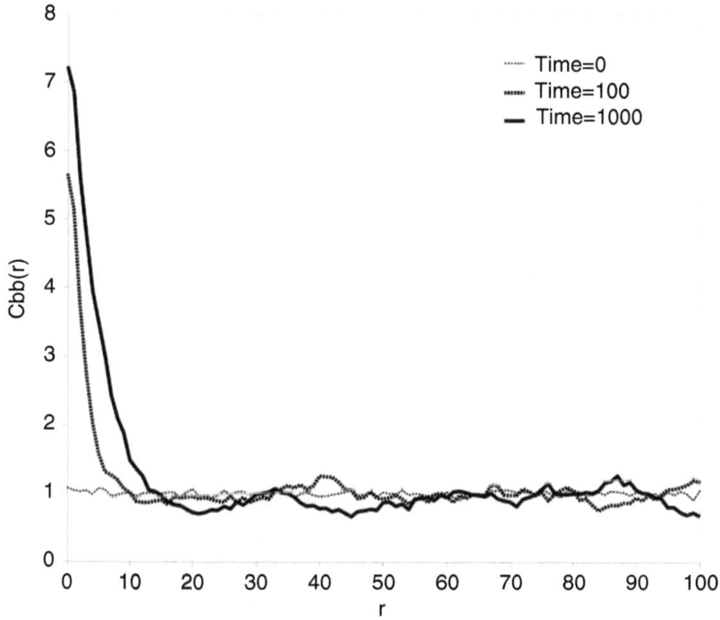

Fig. 6.3 Time evolution of the pair correlation function. The simulation parameters are $b = 0.12$, $d = 0.05$, $d' = 0.0$, $m_1 = 0$, $m_2 = 0$, $w_b = 3$

detachment kernel. The distance between the colonies increases when we increase the size of the detachment kernel (data not shown) These patterns tend to minimize the effect of the detachment. If the colonies are too close to each other (separated by a distance smaller than the radii of the detachment kernel) they would experience a higher detachment rate due to the higher perceived local density of polymer.

The pair correlation function captures the main feature of this pattern as shown in Fig. 6.5. The stationary pair correlation function shows a wavy shape with peaks separated with a constant distance. The peaks indicate that the average density of neighbours increases at regular distances corresponding (approximately) to a multiple of the radii of the detachment kernel. Note that the polymer accumulates only within the colonies. This is due to the assumption that polymer production and detachment occur at the same rate and spatial scale as bacteria birth and detachment processes.

6.3.3 *Motile Bacteria* ($m_1 = 1.0, m_2 = 0$) *with Density-Independent Detachment* $b > d$ *and* $d' > 0$

One can explore how motility affects these spatial patterns by introducing a density-independent motility. Motile bacteria tend to disperse which prevents the formation

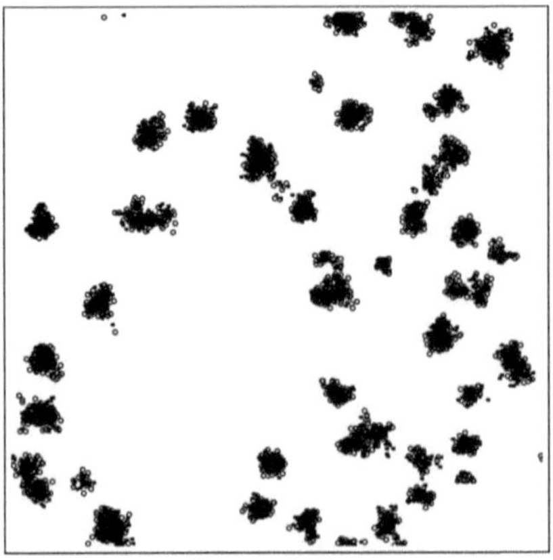

(a) $t = 100$

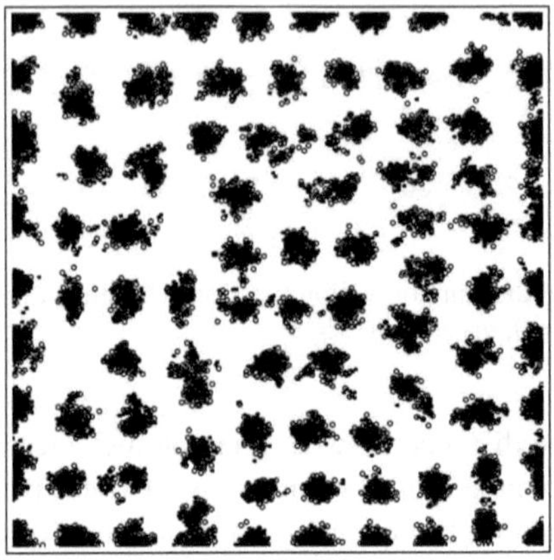

(b) $t = 1000$

Fig. 6.4 Snapshots of the bacteria and polymer spatial patterns. The simulation parameters are $b = 0.12$, $d = 0.05$, $d' = 0.4$, $m_1 = 0$, $m_2 = 0$ $w_b = 3$, $w_d = 31$. *Black* bacteria, *White* polymer

of colonies. Thus, bacteria and polymer are distributed uniformly as shown in Fig. 6.6. The pair correlation function takes values close to 1 confirming that the average neighbourhood of the bacterial cells is close to what would expected from a uniform distribution of the individuals (Fig. 6.7).

6.3.4 Motile Bacteria ($m_1 = 1.0, m_2 = 5.0$) with Density-Independent Detachment $b > d$ and $d' > 0$

We have investigated how the reduction of the bacteria motility due to the production of polymer affects the bacteria spatial pattern. Figure 6.8 shows snapshots of the

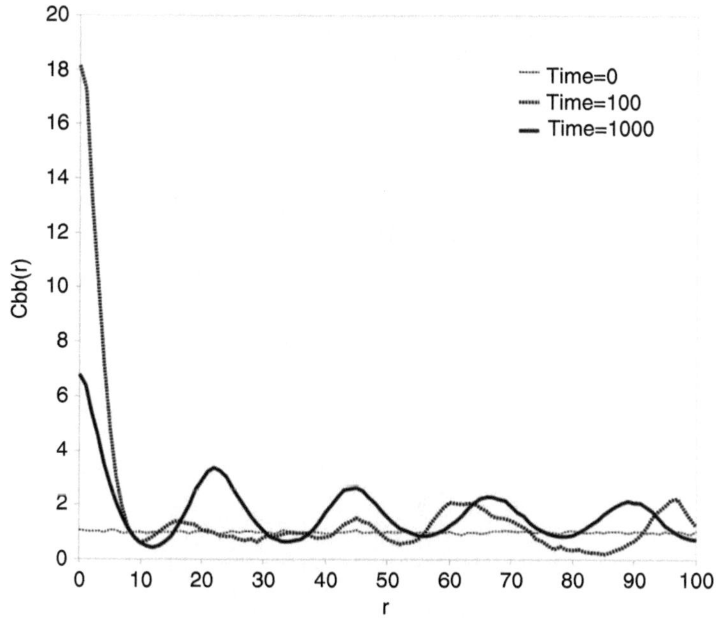

Fig. 6.5 Time evolution of the bacteria–bacteria pair correlation function. The simulation parameters are $b = 0.12$, $d = 0.05$, $d' = 0.0$, $m_1 = 0$, $m_2 = 0$, $w_b = 3$, $w_d = 31$. *Black* bacteria, *white* polymer

transitory and the quasi-stationary spatial pattern. The pattern shows a complex labyrinth-like structure with bands formed by the bacteria and polymer and separated by a regular distance. The bacteria and polymer bands show a structure with cavities in their central parts probably due to high detachment rates in these locations.

6.4 Moment Approximation of the Dynamics

Despite our improved understanding of the mechanisms involved in the biofilm development process, biofilm growth control in still challenging scientists and engineers in several domains including the medical, industrial and environmental fields. Biofilm development is to a wide extent mediated by the local interactions between the individuals and with their local environment. These interactions can be modelled using individual-based models. However, IBMs are difficult to use for addressing control and viability problems because of their complexity, high computational times and stochasticity. To overcome this difficulty we adopt the strategy of approximating the stochastic IBM dynamics with a deterministic aggregated mathematical model.

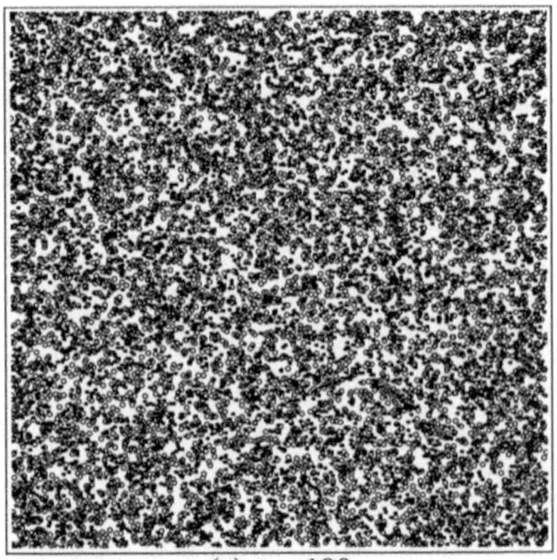

(a) $t = 100$

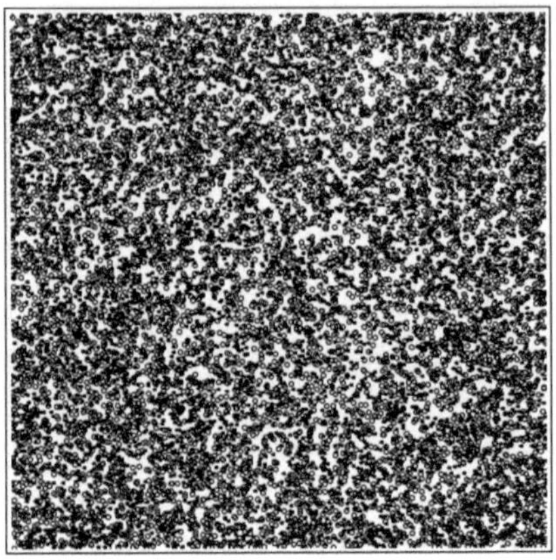

(b) $t = 1000$

Fig. 6.6 Snapshots of the bacteria and polymer spatial patterns. The simulation parameters are $b = 0.12$, $d = 0.05$, $d' = 0.4$, $m_1 = 1.0$, $m_2 = 0$ $w_b = 3$, $w_d = 31$, $w_m = 31$. *Black* bacteria, *white* polymer

Moreover, to facilitate this computation, we considered a simplified version of the previous model, which is now being presented.

6.4.1 Simplifying the Individual-Based Model

The idea to simplify the IBM described above is to drop the polymer dynamics, assuming that bacteria motility and detachment are dependent on the bacteria spatial distribution rather than on the polymer pattern. This simplification implicitly supposes that bacteria and polymer tend to accumulate in the same regions within the domain which is an acceptable assumption if bacteria birth and polymer

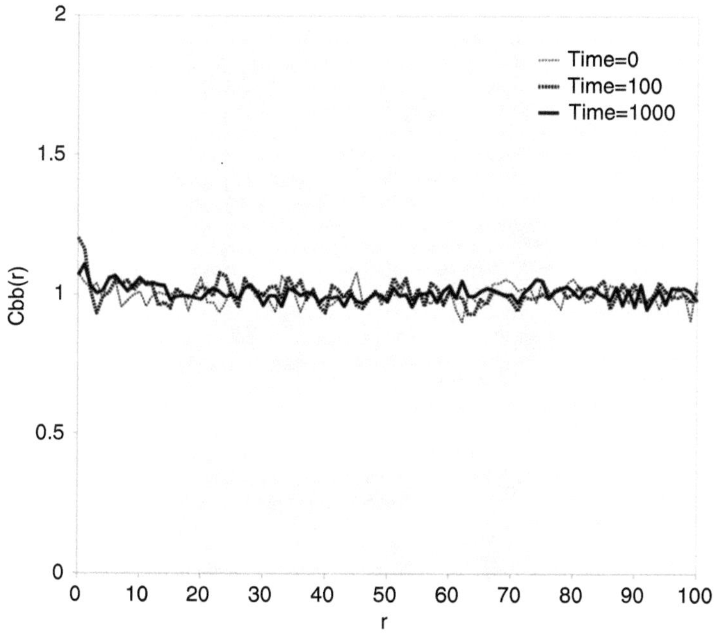

Fig. 6.7 Time evolution of the bacteria–bacteria pair correlation function. The simulation parameters are $b = 0.12$, $d = 0.05$, $d' = 0.4$, $m_1 = 0$, $m_2 = 0$ $w_b = 3.w_d = 31$, $w_m = 31$

production occur at the same spatial scale. We assess through numerical simulations whether this simplified model still allows the obtainment of the spatial patterns identified previously.

Figure 6.9 compares the stationary spatial patterns obtained with the model with polymer and with the simplified model in the case of motile and immotile bacteria. The simplified model allows us to reproduce the spatial pattern that we identified in the detailed model as long as the polymer and bacteria dynamics take place at the same spatial scales.

6.4.2 Moment Model

Classically, the definition of a viability problem requires a dynamical model of the system including at least one controllable variable and a set of constraints on the state of the system (Aubin 1991). In the current problem, we consider that the system is described with the density of bacteria $N(t)$ and the pair correlation density function $C(\xi, t)$, where ξ is a vectorial distance ($\xi = (\xi_1, \xi_2)$). These variables are constrained in the range $[N^L N^U]$ and $[C^L(\xi) C^U(\xi)]$), respectively:

$$\begin{cases} N^L < N(t) < N^U \\ C^L(\xi) < C(\xi, t) < C^U(\xi), \forall \xi \end{cases} \tag{6.11}$$

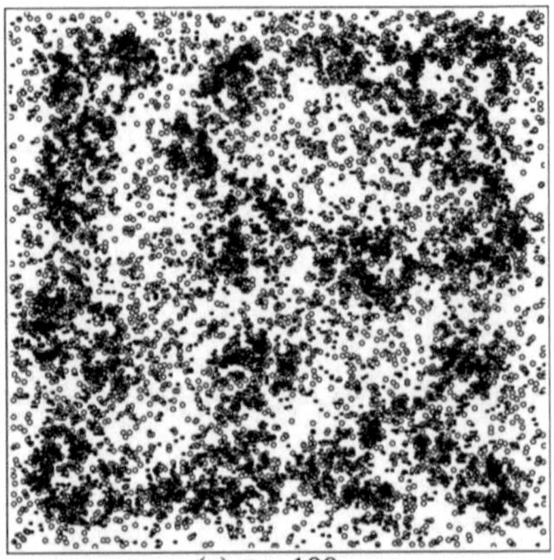

(a) $t = 100$

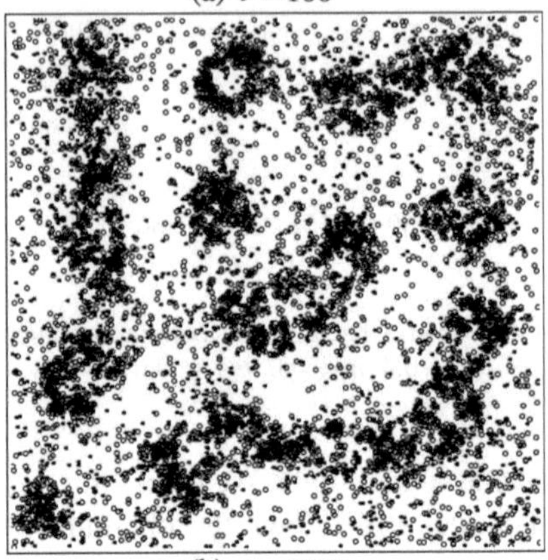

(b) $t = 1000$

Fig. 6.8 Snapshots of the bacteria and polymer spatial patterns. The simulation parameters are $b = 0.15$, $d = 0.05$, $d' = 0.5$, $m_1 = 1.0$, $m_2 = 5.0$, $w_b = 3$, $w_d = 31$, $w_m = 31$, $w_v = 31$. *Black* bacteria, *white* polymer

The dynamics of $N(t)$ and $C(\xi, t)$ can be calculated using the IBM described in the section above. However, the IBM cannot be integrated in classical algorithms for computing resilience because of the stochasticity and the complexity of the model (too many dimensions). IBM require heavy computing resources and hence can only be applied to relatively small populations. Moreover, their results are often difficult to characterize mathematically. It is for instance costly to assess their robustness and to practice sound sensitivity analyses. The idea here is to approximate the dynamics of the IBM and more precisely of the average density and the average pair correlation function measured on the IBM simulations with a deterministic mathematical model. This can be achieved using moment approximation techniques. Dieckmann and Law (2000) showed the interest of the moment approach in approximating the dynamics of several IBMs, in particular IBMs based on the

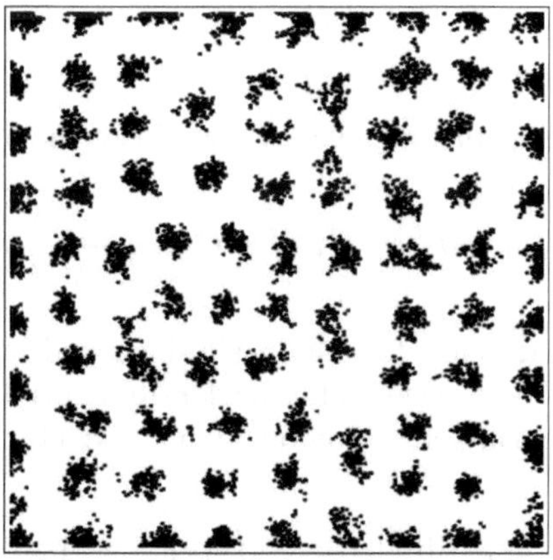

(a) immotile bacteria (equivalent to case 2)

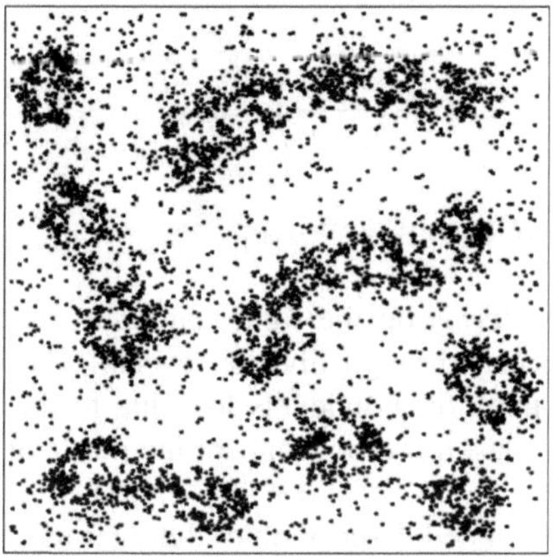

(b) motile bacteria (equivalent to case 4)

Fig. 6.9 Snapshots of the (quasi-) stationary spatial pattern (time = 800) yielded by the simplified model. (**a**) immotile bacteria (equivalent to case 2) (**b**) motile bacteria (equivalent to case 4)

spatial logistic model with birth, death and migration processes. The approximated dynamics are defined with a small number of equations, yet they can capture the main features of complex spatio-temporal phenomena. They can be run with large populations, and their results are easier to interpret mathematically.

We propose to approximate the dynamics of the average density of individuals and the pair correlation function using moment approximation technique. We introduce an additional process that can be controlled and which acts on the spatial pattern. The process consists in adding to the system new individual bacteria cells at a rate c_1. The introduced cells are distributed randomly over the domain. Hence, the state variable of the moment model are the average density of individuals N (the first spatial moment) and the pair correlation function $C(\xi)$ (the second spatial moment) where $\xi = (\xi_1, \xi_2)$ is a vectorial distance and the rate on introduced individuals c_1. We suppose that the rate c_1 can be varied according to a control function u such that:

$$\frac{dc_1}{dt} = u \tag{6.12}$$

The dynamics of the spatial moments N and C is given by:

$$\frac{dN}{dt} = (b_1 - d_1)N - d_1' \int C(\xi) K\left(\frac{||\xi||}{w_d}\right) d\xi + c_1 \tag{6.13}$$

The term $(b_1 - d_1)N$ accounts for the density-independent birth and detachment rates while the integral terms account for the density-dependent detachment.

The dynamics of the pair correlation function is given by:

$$\begin{aligned}\frac{1}{2}\frac{dC(\xi)}{dt} = &\, b_1 N K\left(\frac{||\xi||}{w_b}\right) + b_1 \int K\left(\frac{||\xi'||}{w_b}\right) C(\xi + \xi') d\xi' \\ &- d_1' C(\xi) K\left(\frac{||\xi||}{w_d}\right) - d_1' \int T(\xi, \xi') K\left(\frac{||\xi'||}{w_d}\right) d\xi' \\ &- d_1 C(\xi) + c_1 N + c_1^2\end{aligned} \tag{6.14}$$

The first and second terms on the right-hand side account for the formation of new pairs at distance ξ through division events. The third, fourth and fifth terms account for the pairs of individuals lost due to the detachment of one of the individuals. The two last terms account for the new pairs created by the added individuals.

Following is a more detailed description of these terms:

- The first term accounts for the density-independent division of an individual i producing a new individual j located at a vectorial distance ξ. Multiplying the mean density of individuals N and the independent per capita division rate b_1 gives the rate of division events. Then we multiply by the probability that the newly formed cell is located at distance ξ from the parent position
- The second term also accounts for the density-independent division, but focuses on the new pair that the offspring of an individual i forms with and individual j located at a distance $\xi + \xi'$ from i. The per capita rate of density-independent rate of division is b_1, the density of ij pairs is $C(\xi + \xi')$ and the spatial density of offspring settling around the i parent is $K(||\xi'||/w_b)$. Multiplying these three factors and integrating over all possible distances ξ' of offspring dispersal yields the third term
- The third term accounts for ij pairs that are lost due to the density-dependent death of the individual i in the pair ij due to the presence of the individual j
- The fourth term accounts for ij pairs that are lost due to the density-dependent death of the individual i in the pair ij due to the presence of an individual k at a distance ξ' from the individual i

- The fifth term accounts for ij pairs that are lost due to the density-independent death of the individual i in the pair
- The sixth term accounts for the ij pairs formed with exiting individuals i and the added individuals j
- The seventh term accounts for the ij pairs formed with two added individuals i and j

The factor $^1/_2$ on the left-hand side accounts for newly formed individuals that disperse to distance $-\xi$ which also form a new pair (j,i) at distance ξ. The pair correlation equation involves the third spatial moment $T(\xi,\xi')$ which measures the average density of triplets of individuals with a focal individual separated by a vectorial distance ξ from the second individual and a distance ξ' from the third one. Theoretically, it is possible to derive a dynamical equation for the third moment, but this equation will involve higher order moments. The derivation of the dynamical equations of all the spatial moments makes the resulting moment model of high complexity. Hence in practice the moment cascade of equations is closed at the second moment. This is performed using a closure equation that expresses the third moment as a function of the first and the second moments. There are many possible *closure expressions* (see Dieckmann 2000 for a complete list) and the choice of a particular closure expression is often guided by the quality of the fit between the dynamics calculated by the moment model and simulated with the IBM. We tested the following two closure expressions that we refer to respectively as the power-2 (6.15) and power-3 (6.16) closure expressions:

$$T(\xi,\xi') = \frac{C(\xi)C(\xi')}{N^2} \tag{6.15}$$

$$T(\xi,\xi') = \frac{C(\xi)C(\xi')C(\xi'-\xi)}{N^3} \tag{6.16}$$

We solve the moment model formed with (6.13), (6.15) and (6.15) (or (6.16)) by discretizing the vectorial distances ξ with a spatial resolution $d\xi = (d\xi_1, d\xi_2)$ and time with a constant time step $\Delta t = 1$. We use an explicit Euler scheme for descritizing the time derivative. The resultant algebraic system is formed with n_x^2+1 equation (where $n_x = 101$ is the size of discretized $C(\xi)$ expresses the density of individuals N and the pair correlation matrix $C(\xi)$ at the instant $t+\Delta t$ as a function of N and C at the previous instant t.)

6.4.3 *Comparison of the Moment Model and the IBM Simulation*

The moment model provides a deterministic approximation of the average dynamics of the IBM. The quality of this approximation can be assessed by comparing the dynamics of the average density of individuals yielded by the IBM and the

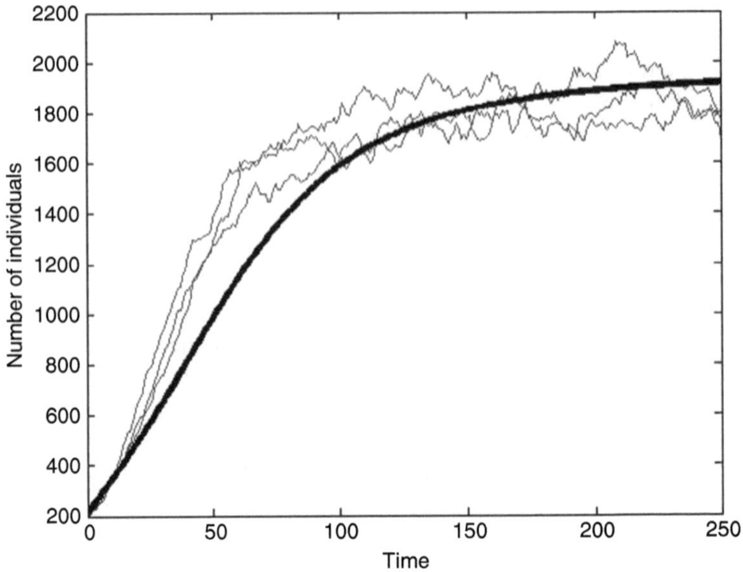

Fig. 6.10 Comparison of the dynamics of bacteria density simulated by the IBM and calculated by the moment model Domain size $= 1 \times 1$, $b_1 = 0.12$, $d_1 = 0.05$, $d_1' = 4.010^{-5}$, $s_b = 0.01$, $s_d = 0.07$, $r_b = 0.03$, $r_d = 0.27$, $N_0 = 200$, $n_x = 101$

moment model (Dieckmann 2000). The choice of the closure expressions can have a significant impact on the quality of the fit between both models. Figure 6.10 shows that the density of individuals (first moment) from the moment model and derived from the IBM are reasonably close.

6.5 Practical Solution of the Viability and Resilience Problem

6.5.1 Reduction of the Moment Model State Space

The state variables of the moment model are the average density of individuals N, the pair correlation function $C(\xi)$ and the feeding rate of individuals c_1. After spatial discretization, the pair correlation matrix is represented with a $n_x \times n_x$ matrix which yields $n_x^2 + 2$ state variables ($n_x = 101$). While the moment model has the advantage of providing a deterministic description of the average IBM dynamics, it still evolves in a large state space to solve the viability problem in reasonable computational times. Hence we propose to project the dynamics of the moment model in a reduced state space. We introduce an aggregated variable (called POS) measuring the regularity of the spatial pattern and calculated from the $C(\xi)$ matrix:

$$POS = \int_{-\infty}^{+\infty} (C(\xi) - \langle C \rangle)\delta_{c(\xi)>\langle C \rangle} d\xi \tag{6.17}$$

In this equation, $\langle C \rangle$ denotes for the average value of the elements of the pair correlation matrix, $\delta_{C(\xi)>\langle C \rangle} = 1$ if $C(\xi) > \langle C \rangle$, and is null otherwise. This pattern regularity indicator sums up the parts of the pair correlation matrix which are higher than the average. Hence the indicator value increases for a wavy pattern showing regular peaks and is close to 0 for a uniform pattern. For a uniform pattern all the elements of the pair correlation matrix have the same value which is also the average value, and this indicator is close to 0. On the contrary, if the pattern shows a lot of oscillations, then the value of the indicator will be high.

In addition to the variable POS we add two other state variables: N^2 and c_1. We selected the variable N^2 instead of N for practical convenience to better highlight (in the reduced space) the small variations of the density of individuals N. The state of the moment model can then be reduced to the three variables (POS, N^2, c_1). Figure 6.11 shows an example of the moment model trajectories in the reduced space for a control $u = 0$ (c_1 is constant) and different initial values of c_1. For $c_1 = 0.0$ (no addition of bacteria cells) the indicator of pattern regularity (POS) increases due to the formation of the wavy pattern (as also shown by the IBM simulations in Fig. 6.12). The density of individuals N (and hence N^2) increases and reaches a stationary value $N^2 = 0.11$. The increase of c_1 yields spatial patterns with a higher number of individuals (higher N^2) and lower values of the pattern regularity indicator as shown in Fig. 6.11. The lower values of the pattern regularity are due to the uniform addition of individuals which tends to flatten the distribution. Note that the dynamics of the moment model is still calculated using the original model involving the state variables c_1, N and $C(\xi)$.

6.5.2 Viability-Resilience Calculation

The viability problem can now be formulated in the reduced space POS, N^2, c_1. These variables are constrained in the range $[POS^L POS^U]$, $[{N^2}^L {N^2}^U]$ and $[c_1^L c_1^U]$) respectively:

$$\begin{cases} POS^L < POS(t) < POS^U \\ {N^2}^L < N^2(t) < {N^2}^U \\ c_1^L < c_1 < c_1^U \end{cases} \tag{6.18}$$

The dynamics of POS, N^2 and c_1 are deduced from the state variables of the moment model. The calculation of the viability kernel requires to discretize the reduced state space into a number of discrete positions. The viability kernel calculation algorithm iterates over these locations and assesses whether they are viable or not (see Chap. 7). This requires to run the moment model by taking each point in the reduced space as an initial point. However, as the state variables of the moment model are given by the vector $(N, C(\xi), c_1)$, we need (in principle) to reconstruct the state vector of the moment model for any point in the reduced space. This is not possible as we cannot ensure that for a point in the reduced space we have a unique corresponding point in the space (N, C, c_1).

To solve this problem we proceed as follows:

- We perform a set of simulations using the moment model starting from different initial conditions to obtain a set of trajectories in the space (N, C)
- We project these trajectories in the plane (N^2, POS) in the reduced space. The obtained set of points defines a two-dimensional sample covering the relevant part of the constraint set and for each point we know the corresponding values of N and C (see Fig. 6.11)
- As c_1 is a common variable for the state space (N, C, c_1) and the reduced space, we replicate the two-dimensional sample in the c_1 dimension to obtain the sample in the reduced space (N^2, POS, c_1) (see Fig. 6.13) Note that we know for each point of the three-dimensional sample (N^2, POS, c_1) the corresponding point in the space (N, C, c_1). The sample in the reduced space can be coarsened by removing the points which are very close to each others. Ideally the trajectories should not intersect to avoid to have for a given point in the reduced space more than one corresponding point in the space (N, C, c_1)

We verified that the description in the space (N^2, POS, c_1) is a good way to characterize the patterns. Globally, increasing c_1 leads to an attractor with a larger N^2, and a lower POS. However the sample does not cover systematically the reduced state space as we have in the other case studies of the book (with a state space discretized using a regular grid). This can lead to a less accurate approximation of the viability kernel.

We describe now examples of viability computing on this complex pattern dynamics. We suppose that we want to control how the bacteria self-organize, by adding a part of them with a controllable rate c_1. Hence we suppose that the control variable is a modification Δc_1 of c_1, that we suppose to be limited between two values: $\Delta_m < \Delta c_1 < \Delta_M$. Therefore, we can add more or less bacteria into the system. We do not claim that the exercise is realistic. Our goal is to demonstrate the possibility of using the viability resilience approach in such a context.

Figure 6.14a shows an example of viability kernel computation, and a corresponding trajectory in the case where an attractor is included in the constraint set. As expected, the trajectories of lazy control (see Chap. 2) lead to this attractor. If the constraint set does not contain any attractor it becomes more difficult to maintain the system, because regular actions should be taken. Figure 6.14b illustrates this case. Note that the fluctuations on the top left may lead to cross the constraints for a short time. This is probably due to the different approximation in the sample and the dynamics.

Considering the same examples, we now study the resilience of the system, in the sense introduced in Chap. 2. The principle is to extend the considered space beyond the constraint set, supposing that the system could get out of it because

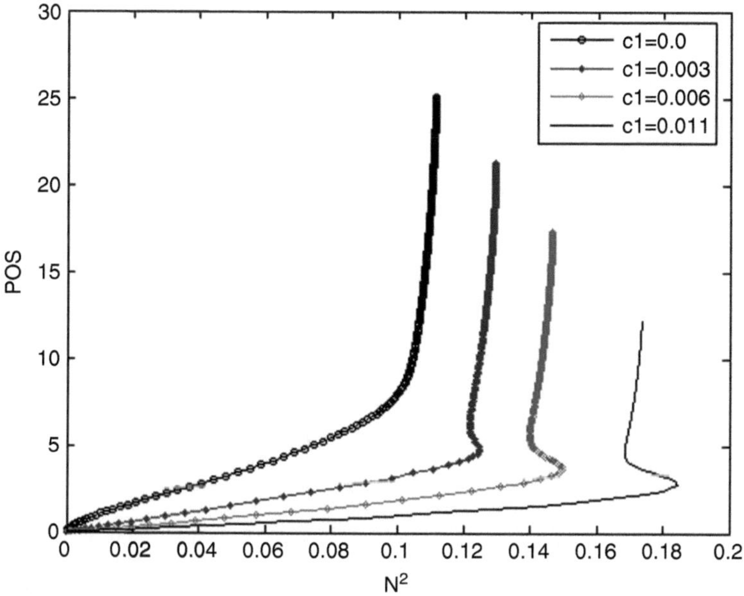

Fig. 6.11 Trajectories of the moment model in the reduced space (N, POS) for different initial values of c_1 (and a control $u = 0$)

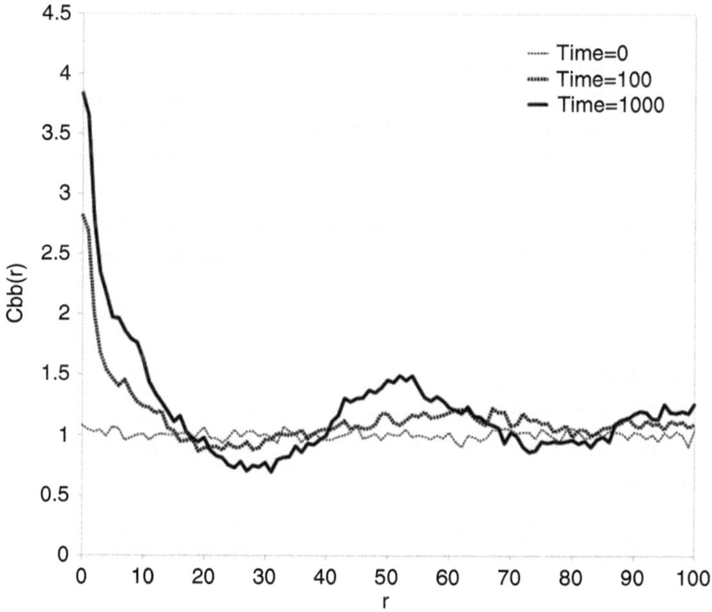

Fig. 6.12 Time evolution of the bacteria–bacteria pair correlation function. The simulation parameters are $b = 0.12$, $d = 0.05$, $d' = 0.0$, $m_1 = 0$, $m_2 = 0$, $w_b = 3$

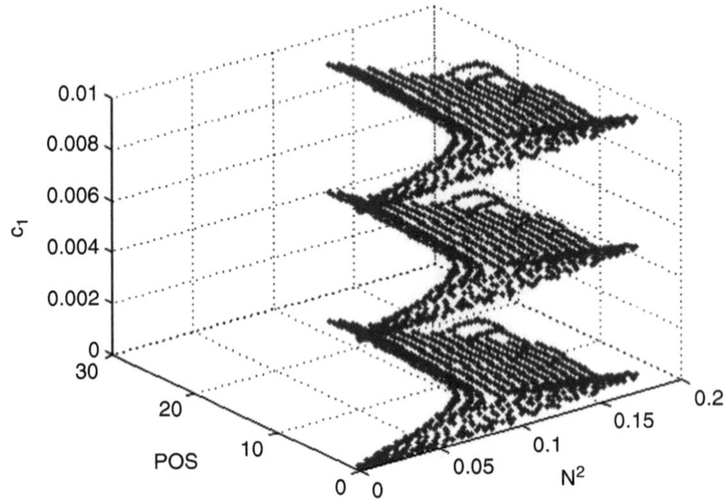

Fig. 6.13 Replication of the trajectories in the space (N, POS) in the c_1 dimension

of a perturbation. In this case, we focused on the part of the space which is higher than the constraint on *POS*, the indicator of regularity. This corresponds to situations where the pattern formed colonies spaced with a regular distance yielding high waves in the correlation matrix. We investigated whether it is possible to modify this pattern into a more uniform pattern where the colonies are less differentiated.

Figure 6.15 shows examples of such calculations corresponding to the cases where an attractor is included in the constraint set, and when none is included. When an attractor is included in the constraint set (Fig. 6.15a) the algorithm computes a set of resilience basins which finally include the whole set. In case where the constraint set does not include any attractor, we note that the resilience basin bigger than the viability kernel is found. It means that in this case, the actions of changing c_1 do not manage to deflect the trajectories sufficiently towards the attractor.

6.6 Conclusion

This chapter provides a novel aspect to the global approach advocated in this book. Its main steps are:

- First, develop a detailed model expressing the details of the system of interest. This model allows us to identify the key processes that drive the system dynamics and the development of complex patterns. In this chapter, this model corresponds to the individual-based model described by Mabrouk (2009) regarding bacteria

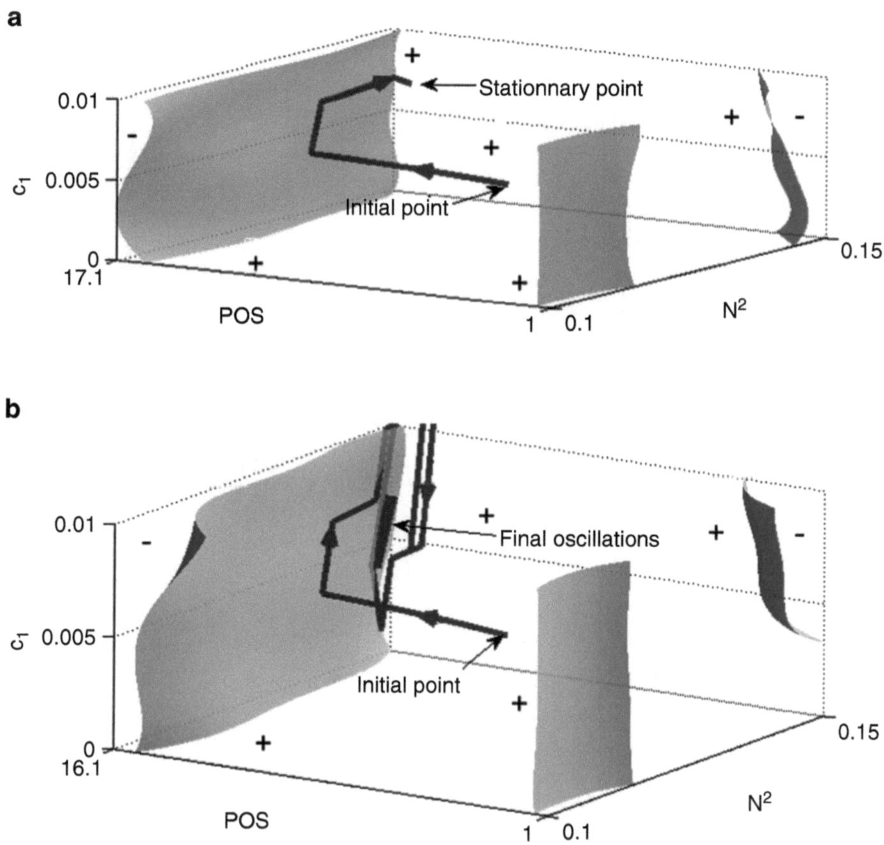

Fig. 6.14 Examples of viability kernels and controlled trajectories. The viability kernel is delimited by the grey surfaces. The (+) sign indicates the viable side. The represented trajectories (*black line*) start from the point $N^2 = 0.13$, $POS = 8$, $c_1 = 0.004$. (**a**) This is the case where an attractor is contained in the viable kernel. In this case the trajectory goes to the attractor (stationary point). (**b**) When the attractor is outside the constraint set, the system has to modify regularly its control to remain in the constraint set. Because of the inaccuracies of the approximations, the trajectory may cross the boundaries of the constraint set

and polymers, competition for space (shoving process) and precise modelling of the substrate dynamics

- Second, simplify the detailed model by keeping only the identified key processes. This may require to design new simpler models which focus more precisely on the complex phenomenon of interest. In this chapter, we were interested in the emergence of spatial structures due to the interplay between birth, detachment and motility processes. Hence, substrate dynamics and the spatial extent of the cells were neglected. We showed that the model can even further be simplified by neglecting the polymer dynamics and approximated with a deterministic moment-based model

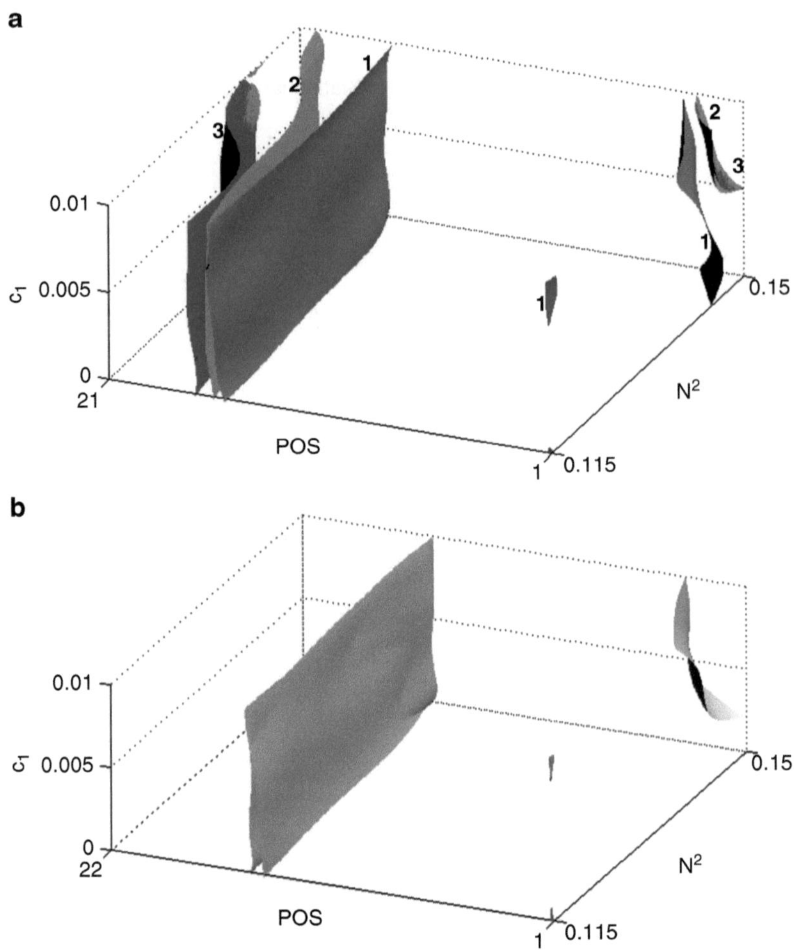

Fig. 6.15 Examples of resilience basins and controlled trajectories. The viability kernel is delimited by the grey surfaces labelled (1). (**a**) There is an attractor in the constraint set. The limits of successive resilience basins are labelled 2 and 3. (**b**) The constraint set contains no attractor. No resilience basin bigger than the viability kernel is found. Once the system quits the viability kernel, there is no way to come back to it

- Finally, address the viability problem on the simplified model. If the state space of the simplified model is still large, extract synthetic indicators that aggregate the state of the system into a smaller number of variables and use them to define the state space for the viability-resilience approach. In this chapter, this indicator is the integral of the correlation matrix over its average, which is found relevant to characterize the pattern regularity

This approach can be adapted to other case studies, but this chapter allowed us to identify several caveats. First, there is of course no guarantee to always find a small number of indicators which summarize the features at stake. It is even probably exceptional that a single indicator is sufficient to characterize the complex shape of a matrix. For instance, this indicator would have certainly been insufficient to treat also the labyrinth-like patterns. Moreover, we have seen in the examples that taking the points from generated trajectories instead of a systematic sampling over the constraint set can lead to some inaccuracies of the standard algorithms for approximating viability kernels and computing a resilience index. It is hence difficult to give strong guarantees about the action policy derived from these computations.

Despite all these weaknesses that should not be underestimated, this chapter shows on an example that it is possible to define policies of actions to favour the maintenance or the recovery of some types of emerging phenomena. In this particular case, it was a level of regularity in the biofilm pattern. We achieved this mainly by making a link between a synthetic space of indicators, which was unequivocally associated with more complex descriptions of the patterns (the correlation matrices). Hence we could compute the complex dynamics of patterns in their natural space, but compute also the viability kernels and resilience indexes in the synthetic space. It seems legitimate to talk of 'pattern resilience' in this case.

References

Aubin JP (1991) Viability theory. Birkhauser, Springer

Costerton J, Lewandowski Z, Caldwell D, Korber D, Lapin-Scott H (1995) Microbial biofilms. Annu Rev Microbiol 49, 711–745

Dieckmann U, Law R, Metz JAJ (2000) The geometry of ecological interactions. Cambridge University Press, Cambridge

Grimson M, Barker G (1994) Continuum model for the spatiotemporal growth of bacterial colonies. Phys Rev E 49, 1680–1684

Hellweger FL, Bucci V (2009) A bunch of tiny individuals–Individual-based modeling for microbes. Ecol Model 220, 8–22

Kreft J-U, Picioreanu C, Wimpenny JWT, van Loosdrecht MCM (2001) Individual-based modelling of biofilms. Microbiology 147, 2879–2912

Luke S, Cioffi-Revilla C, Panait L, Sullivan K (2004) MASON: A Java Multi-Agent Simulation Toolkit. Proceedings of the 2004 SwarmFest Workshop

Mabrouk N, Deffuant G, Tolker-Nielsen T, Lobry C (2009) Bacteria can form interconnected microcolonies when a self-excreted product reduces their motility: Evidence from individual-based model simulations. Theory Biosci. DOI 10.1007/s12064-009-0078-8

Picioreanu C, Kreft J-U, van Loosdrecht MCM (2004) Particle-based multidimensional multi-species biofilm model. Appl Env Microbio 70(5), 3024–3040

Whitchurch C, Tolker-Nielsen T, Ragas PC, Mattick JS (2002) Extracellular DNA required for bacterial biofilm formation. Science 295

Part III
Tools and Techniques

Chapter 7
Approximating Viability Kernels and Resilience Values: Algorithms and Practical Issues Illustrated with KAVIAR Software

Laetitia Chapel and Guillaume Deffuant

7.1 Introduction

In Chap. 2, we presented the definition of resilience based on viability theory (Martin 2004), and we argued that this definition is more general than the equilibrium based definition, and fits better the usual meaning of resilience. Several chapters illustrate this approach on individual based models (language dynamics, bacteria, savanna). In each of these case studies, a preliminary work is to approximate the individual based model with more of a synthetic model, because the tools for computing viability kernels cannot deal with dynamical systems with a state space of high dimension.

In this chapter, we focus on the computation of viability kernels and resilience basins in order to define viable or resilient policies of action. In particular, we explain in more detail how the problem of the state space dimensionality is connected with the well known "dimensionality curse", which appears generally in optimal control problems. To summarise, this difficulty comes from the need to discretise the state space, in order to compute how the dynamical system behaves locally everywhere in the space. From this computation, one can get a global picture by connecting all these local behaviours. But discretising the state space generates a number of points which increases exponentially with the state space dimension. Hence the computation requires exponentially increasing resources in memory and computing power. The limit of current standard computers is very rapidly reached. For instance, in a state space of dimension six, if we build a regular grid of 20 points

L. Chapel
Cemagref - LISC, 24 av. des Landais 63172 Aubière, France

Université de Bretagne Sud, Lab-STICC 8 rue Montaigne, 56017 Vannes Cedex, France
e-mail: laetitia.chapel@univ-ubs.fr

G. Deffuant (✉)
Cemagref - LISC, 24 av. des Landais 63172 Aubière, France
guillaume.deffuant@cemagref.fr

G. Deffuant and N. Gilbert (eds.), *Viability and Resilience of Complex Systems*, Understanding Complex Systems, DOI 10.1007/978-3-642-20423-4_7,

per dimension, then we get 64 million points in total, which begins to be uneasy to manipulate. Moreover, the dimensionality curse related to the dimension of the action (control) space is another challenge in this type of problem. Indeed, in the most basic approach, one has to discretise the space of actions in order to explore it at each local point of the state space. Again, the required computing resources increase exponentially with the dimension of the control space.

This chapter emphasises a particular set of algorithms which approximate viability kernels and resilience basins using statistical classification techniques. This approach has been developed in the PATRES project, using a particular classification technique: Support Vector Machines (SVM). Its main advantages are:

- It provides a parsimonious definition of the viability kernel, in a continuous state space, from which one can define compact and fast controllers
- It gives the possibility to use standard optimisation techniques to find the control, potentially breaking the control space dimensionality curse

Moreover, we derive a specific algorithm to compute resilience basins directly in the state space, whereas the theoretical approach requires to compute a viability kernel in an extended state space, with an additional dimension representing the cost. Practically, the algorithm returns a set of resilience basins, corresponding to increasing restoration costs. From these resilience basins, it is possible to define action policies that drive back the system into the viability kernel.

We illustrate the presentation of these algorithms using KAVIAR (Kernel Approximation for VIAbility and Resilience) which is dedicated software for easily using SVM viability kernels and resilience values approximation algorithms. This chapter complements the KAVIAR user guide, which gives technical tools to use the software, whereas this chapter gives clues to obtain "good" approximations and design "good" controllers. We test the algorithms and software on the Abrams and Strogatz model as an example (see Chap. 3 for more details), taking different parameter values and explaining the results. We show that, when parameters are carefully chosen, accurate and reliable approximations are obtained.

7.2 Using Discrete Space Dynamical System

We first focus on approximating viability kernels and resilience values, which is a prerequisite for computing viable and resilient policies of action. We see how the approximation process suffers the dimensionality curse and then that its application is limited in practice to problems with less than 6 or 7 dimensions.

7.2.1 The Viability Algorithm

Several algorithms have been devised to approximate the viability kernel of a system. The seminal one, the so-called "viability algorithm" (Saint-Pierre 1994), is based on a time and space discretisation. In this first section, we describe this algorithm because the SVM viability kernel approximation is based on it.

Consider the following continuous dynamical system, defined by its state variables $x(t)$ along time $t \in \mathbb{R}^+$, and suppose that its evolution can be modified by control variables $u(t)$:

$$\begin{cases} x'(t) = \varphi(x(t), u(t)) \text{ with } u(t) \in U(x(t)) \\ x(0) = x_0 \in K \end{cases} \tag{7.1}$$

where $x(t) \in \mathcal{X} \subset \mathbb{R}^n$ is a vector and K is a subset of $\mathcal{X}$. The action should be chosen in a set $U(x) \subset \mathbb{R}^q$. The system's dynamic is then $\varphi : \mathcal{X} \times U \to \mathcal{X}$.

Saint-Pierre 1994's algorithm provides an approximation of the viability kernel K $Viab_\varphi(K)$ when φ satisfies some conditions.[1] He defines a sequence of discrete viability kernels using discrete approximations of the dynamical system (7.1). He shows that this sequence converges to the viability kernel if $\varphi(x, u)$ is Lipschitz.[2]

The idea of the algorithm is first to discretise the dynamical system in time, choosing a time step dt and using the Euler approximation of system (7.1):

$$\begin{cases} x(t + dt) = x(t) + \varphi(x(t), u(t))dt = G(x(t), u(t)) \\ u(t) \in U(x(t)). \end{cases} \tag{7.2}$$

Then he discretises the state space, defining a grid of points K_h of resolution h covering K such that:

$$\forall x \in K, \exists x_i \in K_h \text{ such that } \|x - x_i\| \leq h. \tag{7.3}$$

He uses this discrete set to define a discrete dynamical system in space and time, approximating the dynamical system defined by G. This discrete dynamical system is defined by function $G_h : K_h \times U \to K_h$:

$$G_h(x_i, u_i) = \{x_j\} = \arg\min_{x_k \in K_h} \|x_k - G(x_i, u_i) \cdot \mu \cdot \boldsymbol{B}\| \tag{7.4}$$

given a control $u_i \in U(x_i)$, with φ μ-Lipschitz and $\boldsymbol{B}$ a ball of radius 1 and centre 0. It then requires the discretisation of the control space.

The algorithm uses the property that the viability kernel $Viab_{G_h}(K_h)$ coincides with the biggest set E such that from any point x in E there exists an action u with $G_h(x, u) \in E$.

Hence, Saint-Pierre considers the sequence $(K_h^n)_n$, with $K_h^0 = K_h$:

$$K_h^{n+1} = \{x_k \in K_h^n, \exists u \in U(x_k) \text{ such that } G_h(x_k, u) \in K_h^n\}, \tag{7.5}$$

[1]The associated set valued map F is Marchaud: it is upper semicontinuous, with compact convex values and verifies $|F(x)| = \sup\{|y| \, |y \in F(x)\} \leq c(1 + |x|), \forall x \in \mathbb{R}^n$.

[2]φ is μ-Lipschitz if there exists one constant μ such that $\forall x, y \in \mathcal{X}, \exists (u, v) \in U(x) \times U(y)$ such that $\|\varphi(x, u) - \varphi(y, v)\| \leq \mu \|x - y\|$).

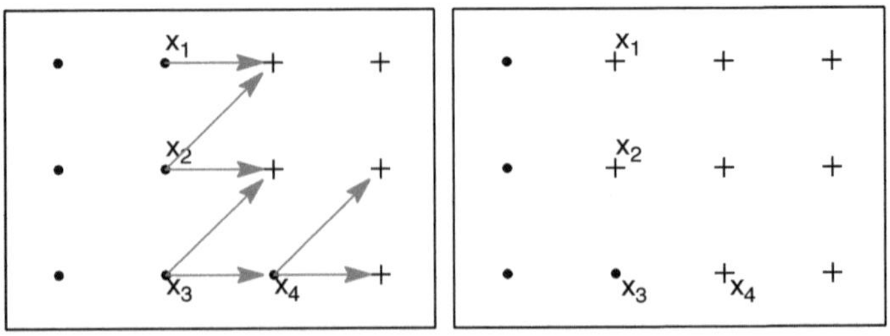

Fig. 7.1 Example of a progressive approximation, from iteration n (*left*) to iteration $n+1$ (*right*). Sets K_h^n and K_h^{n+1} are the set of viable points at each iteration and are represented by *black discs*. Non-viable points are represented by a '+' sign. The successors (for different possibilities of control) of a few points are designated by *arrows*. Viable points which have no viable successors become non-viable (x_1, x_2 and x_4) at the next iteration, while x_3 remains viable as it has (at least) one viable successor

It can be shown that $Viab_{G_h}(K_h) = \bigcap_{n=0}^{+\infty} K_h^n$ and there exists one integer p such that $K_h^p = Viab_{G_h}(K_h)$.

This sequence gives one iterative algorithm to approximate viability kernels by using a discrete approximation: from K_h^n, the algorithm deletes at iteration $n+1$ the points of the grid that do not have any successors in K_h^n. Figure 7.1 gives an example of a progressive approximation, from iteration n to iteration $n+1$.

Moreover, Saint-Pierre shows that, when φ is μ-Lipschitz, the discrete viability kernel tends to the viability kernel of the initial system when the resolution of the grid h tends to 0.

The algorithm is very fast, but at each iteration, the current approximated viability kernel K_h^n is defined as a set of points, which is not very convenient to manipulate. In addition, the size of the grid exponentially increases with the dimension of the state space and the algorithm requires the exhaustive test of all the controls (or the set of the discretised controls when the controls are continuous) – it suffers the dimensionality curse. These two aspects limit the practical use of the algorithm to systems in low dimension, in the state space and in the control space. In addition, the numerical scheme is diffusive (Bokanowski et al. 2006), which gives overestimations of actual kernels.

7.2.2 *Computing Resilience Values*

Computing resilience values comes down to approximating the viability kernel of a new dynamical system with an additional dimension representing the cumulated cost.

First, we introduce the cost function $\gamma(x, u)$ that associates a cost to a point x of the state space when applying action u, and function $C_K(x_0)$ that associates to x the minimal cost on all trajectories starting at x_0:

$$C_K(x_0) = \inf_{u(.)} \int_0^\infty \gamma(x(t), u(t))dt. \tag{7.6}$$

Moreover, we suppose that the cost of being in the viability kernel is null, because by hypothesis, this is the set of states that is desired. We also suppose that being outside the viability kernel has a cost because from the states outside the viability kernel, we are sure that no action policy prevents them from violating the constraints after some time:

$$\gamma(x, u) \begin{cases} > 0 \text{ when } x \notin \mathit{Viab}(K) \\ = 0 \text{ when } x \in \mathit{Viab}(K). \end{cases} \tag{7.7}$$

The resilience problem is thus to determine action policies driving back the system into the viability kernel, at minimum cost. This is a typical optimal control problem. Yet, it can be shown (Aubin and Frankowska 1996) that viability theory can be used to solve it. One needs to extend the state space by adding an axis for the cumulated cost over a trajectory. This new state space includes couples (x, c), where x is in the initial state space, and c represents the cost associated with this point. The constraint set of the extended viability problem is $x \in K$ and $c \geq 0$. The dynamics of the system in discrete time are given by:

$$\begin{cases} (x(t+dt), c(t+dt)) = (x(t) + \varphi(x(t), u(t))dt, c(t) - \gamma(x(t), u(t))dt) \\ u(t) \in U(x(t)). \end{cases} \tag{7.8}$$

It can be shown that the boundary of the viability kernel obtained from the resolution of this problem provides the function which associates the minimum cost $C_K(x)$ to each state x. Resilience values $R_K(x)$ are then defined as the inverse of the minimal cost function to get back to the viability kernel. They correspond to the inverse of the minimal cost of restoration, among all the possible evolutions $x(.)$, if one disturbance brings the system to state x:

$$R_{K(x)} = \frac{1}{C_K(x)}. \tag{7.9}$$

These results imply that resilience values can be computed or estimated using the algorithmic tools used for the viability kernel approximation. But as they require the addition of one dimension (the cost), the dimensionality curse is even more critical than in the viability kernel approximation case.

7.2.3 *Viability and Dimensionality*

There are practical difficulties raised by the increase of the dimension, that are a consequence of what Bellman (1961) coined the curse of dimensionality. In the case of viability algorithms, the dimensionality curse takes place in the state space and in the control space.

It is well known that the number of points of a regular grid covering the unit hypercube grows exponentially with the number of dimensions of the space. Considering a d_s-dimensional unit hypercube, the number of points needed to cover it with a regular grid of step ϵ_s grows as $O((1/\epsilon_s)^{d_s})$, and for each point, $O((1/\epsilon_u)^{d_u})$ computations must be performed (where d_u is the dimension of the control variable, and ϵ_u is the step of the discretisation), that is to say $O((1/\epsilon_s)^{d_s} \times (1/\epsilon_u)^{d_u})$ computations. Consider a 10-dimensional state and control space problem and $\epsilon_s = \epsilon_u = 1/10$ steps, the total number of evaluations to perform is $10^{10} \times 10^{10} = 10^{20}$.

Moreover, given a grid step ϵ, the maximum distance between the grid and the points of the space also increases with d. Indeed, if we consider a unit hypercube, the volume stays the same with the increase of the dimension but the length of the main diagonal grows as $\sqrt{d}$. Taking one random point x on the hypercube and designing the grid such that its maximal distance to the nearest neighbour on the grid G is lower or equal a given value v, then the discretisation step must be equal to $\epsilon = \frac{v}{2\sqrt{d}}$. In order to have any point of the space at a distance lower or equal than 0.05 to the grid, the discretisation step must be equal to $v = 0.1$ in dimension 1 but $v \approx 0.03$ in dimension 10.

In practice, it is generally difficult to manage problems of more than a half-dozen dimensions, using current standard computers. This explains why, in the case studies described in this book, an important part is devoted to the derivation of synthetic models from complex individual based models, in order to compute their viability and resilience.

7.3 Using Classification Procedure and Dynamics in a Continuous Space

7.3.1 *Viability Kernel Approximation Using a Classification Procedure*

Classification procedures are machine learning techniques for deducing a function from a training set of pairs <point,label>, in order to predict a class label of the points. The main motivation behind introducing them is to get a more compact definition of viability kernels and resilience basins. Indeed, the classification procedures generally define a boundary in the continuous space, separating points classified -1 from points classified $+1$, which is easier to manipulate than the explicit set of points (which can be huge). Moreover, we shall see that such a continuous boundary can be used in the process of finding an action that keeps the dynamical system within the current approximation of the viability kernel.

The algorithm is based on the viability algorithm and works roughly as follows: at each iteration, we use a classification procedure to define a continuous approximation of the discrete set K_h^n and we remove only the points of the grid which leave this approximation by more than a threshold amount. The other main difference with the viability algorithm is that we discretise the dynamical system in time but not in space.

7.3.1.1 Definitions and Notations

The classification procedure associates to each point x_k of the grid K_h a label $y_k = +1$ if x_k is viable at the next iteration (i.e. there exists at least one control u that keeps the system inside the current approximation of the viability kernel) and -1 otherwise. From a learning set $\mathscr{S} = \{(x_k, y_k)\}_{x_k \in K_h}$, we define a learner $\mathscr{A} : \mathscr{S} \to l^n(x)$ where the $l^n(x)$ function associates to any state $x \in K$ a label y at each iteration n: $l^n : K \to \{-1; +1\}$. Moreover, we define:

- $d(E, F) = \inf\{d(e, f)/(e, f) \in (E \times F)\}$ the distance between two closed sets E and F
- $E \backslash F$ the complementary set of F in E (supposing that $F \subset E$)
- $L(K_h^{n+1}) = \{x \in K \text{ such that } l^{n+1}(x) = +1\}$
- $\lambda \geq 1$ is a real number

We also assume that we can determine if $\exists u \in U(x_k)$ such that the point $x_k + \varphi(x_k, u)dt$ remains inside the current approximation of the viability kernel.

7.3.1.2 Algorithm

We iteratively define the sets L_h^n such that:

$$\begin{aligned} K_h^0 &= K_h \\ L_h^0 &= K \\ K_h^{n+1} &= \{x_k \in K_h^n \text{ such that } \exists u \in U(x_k), d(x_k + \varphi(x_k, u)dt, L_h^n) < \mu h\} \\ L_h^{n+1} &= L(K_h^{n+1}) \end{aligned} \tag{7.10}$$

If the classification procedure respects the following conditions:

$$\forall x \in L_h^n, d(x, K_h^n) \leq \lambda h, \tag{7.11}$$

$$\forall x \in K \backslash L_h^n, d(x, K_h \backslash K_h^n) \geq h, \tag{7.12}$$

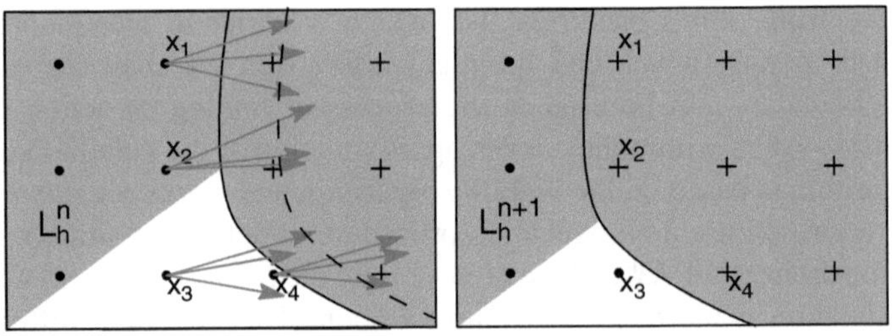

Fig. 7.2 Example of a progressive approximation from iteration n (*left*) to iteration $n+1$ (*right*). The limit of current viability kernel approximations L_h^n and L_h^{n+1} are depicted as a plain line. Sets K_h^n and K_h^{n+1} are represented by black discs and sets $K_h \backslash K_h^n$ and $K_h \backslash K_h^{n+1}$ by '+' signs. Sets L_h^n and L_h^{n+1} are derived from sets K_h^n and K_h^{n+1} thanks to a classification procedure. At iteration n, we test for all the points $x_k \in K_h^n$ if $d(x_k + \varphi(x_k, u)dt), L_h^n) \leq \mu h$ for all $u \in U(x_k)$. Here, to simplify, we supposed that only three values of u are possible. The limit of distance is represented by a *dotted line*. If yes, then $x_k \in K_h^{n+1}$ (x_3 for instance on the figure), otherwise, $x_k \notin K_h^{n+1}$ (x_1, x_2 and x_4 for instance)

then, it can be shown that there exists an integer p which is the first value of n such that $K_h^{n+1} = K_h^n = K_h^p$, and:

$$L_h^p \to Viab_G(K) \text{ when } h \to 0. \tag{7.13}$$

The convergence proofs of this algorithm are given in Deffuant et al. (2007). Figure 7.2 illustrates the passage from iteration n to iteration $n+1$.

7.3.1.3 Viability Controllers

Viable trajectories are those lying entirely within the viability kernel and we know that there exists at least one viable trajectory starting from any viable initial state (Aubin 1991). Aubin also proposes to select among all the viable trajectories the heavy one: the principle is to maintain the controls inherited from the past as long as viability is not at stake, and to apply a control that pushes the trajectory back into the viability kernel otherwise (we know that such a control exists, as long as the considered point lies inside the viability kernel).

The viability controller defined in the case where we have an approximation defined by a classification method adapts the heavy control procedure by adding a security distance to the viability kernel approximation boundary ∂L_h^p. Indeed, we know that the actual kernel is included in the approximation, and we then have to choose a security distance that will allow us to always remain, not only inside the approximation, but inside the actual viability kernel. In addition, instead of looking for the control with minimal velocity when viability is at stake, we pick the one which maximises the distance between the resulting state and the approximation boundary (this aspect is considered in more detail in Chap. 8).

We define parameter $\Delta \in \mathbb{R}^+$ to tune the security distance. We call the algebraic distance from a point to the boundary, denoted $d(x, \partial L_h^p)$, the distance to the boundary with a positive sign if $x \in L_h^p$ and a negative sign otherwise. The following procedure associates a control u_{n+1} at the $(n+1)^{\text{th}}$ iteration:

Heavy controller algorithm

Starting from an initial state x_0 such that $d(x_0, \partial L_h^p) > \Delta$ and a randomly chosen initial control $u_0 \in U(x_0)$, the procedure associates a control u_{n+1} at iteration $n+1$ as follows:

- If $d((x_n + \varphi(x_n, u_n)dt), \partial L_h^p) > \Delta$, we keep the same control $u_{n+1} = u_n$
- Otherwise, $u_{n+1} = \arg \max_{u \in U(x)} \{d((x_n + \varphi(x_n, u)dt), \partial L_h^p)\}$

In practice, there is no proof that this procedure provides a viable trajectory. But under some mild conditions (see Deffuant et al. (2007)), the algorithm will find appropriate controls.

7.3.2 Resilience Basins Approximation and Optimal Controllers Using a Classification Procedure

7.3.2.1 Motivation

As explained before, approximating resilience basins comes down to a viability kernel approximation for a new dynamical system with an additional dimension representing the restoration cost (7.8). But as shown above, viability algorithms face the dimensionality curse and their application is even more restricted in the resilience case, because of this additional dimension. In this subsection, we describe an algorithm which makes all the computations in the initial state space, thus saving one dimension compared with the standard approach. Moreover, this algorithm provides an outer and an inner approximation, which can be used for designing a guaranteed controller procedure.

The algorithm involves two steps:

1. Approximation of the viability kernel on K
2. Approximation of the minimum cost for any point outside *Viab*(K) to go inside $Viab(K)$ (this gives directly the resilience values, as explained before)

To achieve the second step, we design an algorithm which draws its inspiration from the SVM capture basin algorithm in the initial state space, which solves target hitting problems (Chapel and Deffuant 2010). Here, the viability kernel approximated on the first step is the target to reach. When the cost is defined as the time of restoration, this algorithm applies directly. We extend it to other cost functions.

In order to work on the initial state space and define an algorithm inspired from the capture basin algorithm, the cost function γ must satisfy one condition: it must be strictly positive anywhere outside the viability kernel (the target to reach here).

Cost function (7.8) satisfies this property. If, for some reason, the cost function does not satisfy this property, then it is not possible to use the following approach, and the problem is solved by approximating the viability kernel of an extended system, with the cost as an additional dimension.

7.3.2.2 Notations

In order to adapt the capture basin algorithm to the case of a positive cost, which can vary, we have to set a fixed cost variation *dc* for all the states and iterations, and define the dt step accordingly: $dt_{x,u} = \frac{dc}{\gamma(x(t),u(t))}$. We thus discretise the dynamics as follows, for $x \notin Viab(K)$:

$$\begin{aligned}(x(t + dt_{x,u}), c(t + dt_{x,u})) &= (x(t) + \varphi(x(t), u(t))dt_{x,u}, c - dc) \\ &= G_C(x(t), c(t), u(t)).\end{aligned} \tag{7.14}$$

We suppose that G_C is μ_C-Lipschitz. Equation (7.14) amounts to the formulation of a capture basin problem (Aubin 2001) where the time variation dt is replaced by the cost variation *dc* and the target is the viability kernel. Note that we can directly use the capture basin approximation if the cost is the time of restoration ($\gamma(x, u) = 1$ everywhere).

7.3.2.3 Algorithm Using a Classification Procedure

With the same motivation, we derived a version of this approach using a classification procedure. The principles are very similar to the ones already described for viability kernel approximation. The main difference is that we start by approximating the viability kernel and then, gradually add the points that can reach the viability kernel with a cost $c \leq n \times dc$, where n is an integer, instead of removing points as done in viability kernel approximation. This major difference allows us to define an approximation from inside of the same cost surfaces, whereas the approximation of the viability kernel is always from outside (when "good" parameters have been chosen to obtain the approximation). This difference is important, because the approximation from inside defines action policies that guarantee to reach the approximated viability kernel, whereas this is not the case with the approximation from outside.

Let us consider a grid H_h as a finite set of elements of H representing the range of the system state space under study, including the subset of states showing the property of interest $K \subset H$:

$$\forall x \in H, \exists x_k \in H_h \text{ such that } \|x - x_h\| < h, \tag{7.15}$$

At each iteration n, we define a discrete set H_h^n such that $H_h^{n-1} \subset H_h^n \subset H_h$ and a continuous one C_h^n which is a generalization of the discrete set and that represents the cost function values lower or equal than $n.dc$. This continuous set is obtained by training an SVM on the learning sample obtained with the points x_h of the grid H_h, associated with label +1 if $x_h \in C_h^n$ and with label −1 otherwise. Let l_h^{n+1} be the obtained classification function from H to $\{-1, 1\}$.

The steps of the approximation algorithm are the following:

Algorithm

We iteratively define the sets C_h^n such that:

$$\begin{array}{ll} C_h^0 &= L_h^p \\ & \textit{outer approx:} \\ H_h^{n+1} &= \left\{x_k \in H_h \text{ s. t. } \exists u \in U(x_k), d(x_k + \varphi(x_k, u)dt_{x_k,u}, C_h^n) \leq \mu_C h\right\} \\ & \textit{inner approx:} \\ H_h^{n+1} &= \left\{x_h \in H_h \text{ s. t. } \exists u \in U(x_k), d(x_k + \varphi(x_k, u)dt_{x_k,u}, H \backslash C_h^n) > \mu_C h\right\} \\ C_h^{n+1} &= L(H_h^{n+1}) = \left\{x \in H \text{ such that } l_h^{n+1} = +1\right\} \end{array} \tag{7.16}$$

until $H_h^{n+1} = H_h^n = H_h^q$.

If the classification procedure respects some conditions similar to the viability kernel approximation case, and if the dc value is well-chosen (see Chapel and Deffuant (2010)), we have:

$$C_h^q \to \mathscr{C} \text{ when } h \to 0, \tag{7.17}$$

where $\mathscr{C}$ are the actual cost values.

Figure 7.3 illustrates the evolution of the approximation from iteration n to iteration $n + 1$.

7.3.2.4 Optimal Controllers

Resilience values are defined as the inverse of the maximal cost of restoration. Therefore the problem is no longer to find one trajectory among the set of the viable ones but to find a sequence of controls that will drive back the system to the viability constraint set with the minimal cost (and then obtain the maximal resilience). We use the successive resilience basin approximations provided by the algorithm to define an optimal controller.

The aim of the optimal controller in the initial state space is to choose a control series that drives the system back to the approximated viability kernel, without leaving the extended constraint set H. The idea of the procedure is to choose the control that makes the trajectory cross the resilience basins C_h^n in a decreasing order, by following the steepest descent direction.

We start with an initial point $x_0 \in C_h^q$. For each state x_i of the trajectory, we compute $n(i)$ the maximal value of n such that $x_i \in C_h^{n(i)}$. If $n(i) > 0$, we define u_i^* as follows:

$$u_i^* = \arg\max_{u \in U(x)} \left\{d((x_i + \varphi(x_i, u)dt_{x,u}), \partial C_h^{n(i)-1})\right\} \tag{7.18}$$

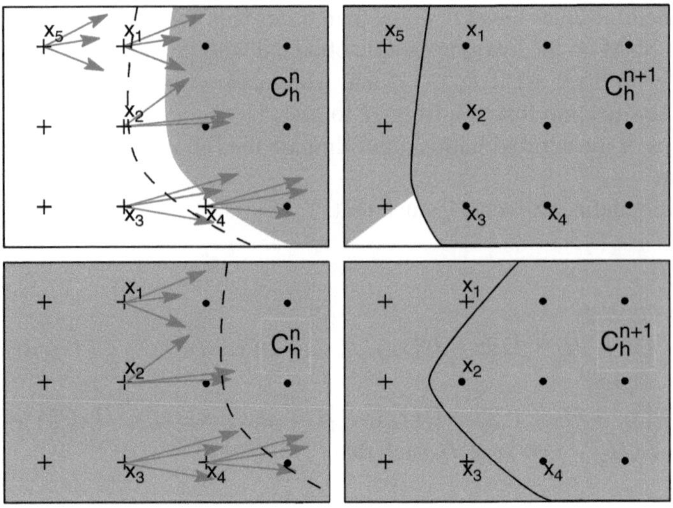

Fig. 7.3 Example of evolution from resilience basin n (on the *left*) to resilience basin $n+1$ (on the *right*): on the *top*, for an outer approximation, on the *bottom*, inner approximation. The resilience basin approximations C^n and C^{n+1} are depicted as *continuous lines*. Points of sets H_h^n and H_h^{n+1} are represented by *black discs* while the points of sets $H_h \backslash H_h^n$ et $H_h \backslash H_h^{n+1}$ are represented by a '+' sign. At iteration n, we test for the points $x_k \in H_h$ if there exists $u \in U(x_k)$ such that $d(x_k + \varphi(x,u)dt_{x,u}, C_h^n) \leq \mu_C h$ (outer approximation) or if $d(x_k + \varphi(x,u)dt_{x,u}, H \backslash C_h^n) > \mu_C h$ (inner approximation). If yes, then $x_k \in C_h^{n+1}$, and $x_k \notin C_h^{n+1}$ otherwise

7.4 Support Vector Machines as a Particular Classification Procedure

7.4.1 Support Vector Machines: A Short Presentation

Support Vector Machines (SVM) are machine learning techniques used for classification or regression. We consider here only the binary classification task: given a set of data points with associated labels $+1$ or -1. Here we provide the minimum to be known for understanding the use of SVMs in the algorithms of viability and resilience. For more details, one can find review books on the topic (Scholkopf and Smola 2002; Cristianini and Shawe-Taylor 2000).

Consider a learning set $\mathscr{S} = \{(x_i, y_i)\}_{i=1}^N$ where x_i is a vector in a space $\mathscr{X} \in \mathbb{R}^d$ and $y_i \in \{-1, +1\}$ is the label associated to x_i. The SVM procedure provides α_i coefficients and scalar value b, of the following non-linear function (also called SVM function):

$$f(x) = \sum_{i=0}^{N} \alpha_i y_i k(x, x_i) + b. \tag{7.19}$$

Function f is used to attribute a label $l(x)$ to any point of the space:

$$f(x) < 0, \Rightarrow l(x) = -1. \tag{7.20}$$

$$f(x) \geq 0, \Rightarrow l(x) = +1. \tag{7.21}$$

The function $k(x, y)$ in (7.19) is called a kernel function. Several kernel functions are commonly used. In the following, we use the gaussian kernel:

$$k(x, y) = \exp\left(-\frac{\|x - y\|^2}{2\sigma^2}\right) \tag{7.22}$$

whose advantage is that the produced SVM function can approach any classification function, thanks to the $\sigma > 0$ parameter, which tunes the standard deviation of the gaussian. By decreasing the σ value, we can approach very irregular shapes while a large σ value produces smoother shapes.

The α are solution of a quadratic problem which is such that generally many α_i values are 0. Examples such that $\alpha_i > 0$ are called support vectors (SV). The fact that function f is defined with a low number of support vectors is connected with the parsimony property of SVMs, which is also linked with their generally good performances as learning techniques.

The computational complexity for solving the SVM problem is $O(N^3)$ with a memory space in $O(N^2)$. Several algorithms have been designed to solve SVM in large dimension. Among them is the Sequential Minimal Optimization algorithm (Platt 1999) which only considers two points at each step and uses shrinking techniques to reduce the needed memory space to $O(N)$.

7.4.2 Approximating Viability Kernels and Computing Viable Action Policies with SVM

In this section, we consider SVM as a particularly relevant classification procedure to approximate viability kernels. The fulfilment of the convergence conditions of the algorithm are discussed in Deffuant et al. (2007). We give the main characteristics of the algorithm for approximating viability kernels, and then show how viable policies of actions can be computed.

We use SVM in order to define functions $l(x)$ and L_h^n used in algorithm (7.10). There are three main advantages of considering SVM as a classification procedure:

- It allows the use of standard optimisation techniques to find a viable control (if any) instead of discretising the set of controls. It then allows one to deal with problems in high dimensional control space
- It provides fast and compact controllers

- It allows one to consider several time steps at each iteration, and then to have a finer approximation of the dynamics

Moreover, we can expect to save memory space, by using only a part of the points of the grid to train the SVM (Chapel and Deffuant 2007).

7.4.2.1 Using an Optimisation Procedure to Find Controls

The analytical expression of L_h^n can be used to find one control (if any) that allows the system to remain inside the current approximation of the viability kernel. Close to the boundary of L_h^n, directions where the SVM function $f_n(x)$ increases go inside L_h^n while those where $f_n(x)$ decreases go toward the outside of L_h^n. $f_n(x)$ then provides a kind of barrier function on which we can build optimisation techniques to find a viable control. In this context, maximising $f_n(x)$ provides a control that keeps the system inside the current approximation L_h^n, if any.

7.4.2.2 Considering Several Time Steps

The advantage of using several time steps at each iteration is that it gives a better approximation of the continuous dynamics. For a given discretisation step, the quality of the time approximation of φ depends on the used scheme: the smaller the dt value, the better the trajectory associated to a point relatively to the continuous dynamics. Indeed, we approach the point $x(t + dt)$ by using a Euler scheme $x(t + dt) = x(t) + \varphi(x(t), u(t)dt$ and the approximation error is in $O(dt^2)$. The dynamics approximation can be improved by using j time steps to compute the successor of a point. For a given value dt, the successor at time $j.dt$ has an error in $O(j.(dt^2))$ instead of $O((j.dt)^2)$ by considering only one time step $dt' = j.dt$.

We can define a series of j optimal controls with a reasonable computational time. First consider an iteratively defined trajectory issued from a point x:

$$\begin{cases} t(x, u_1) = x + \varphi(x, u_1)dt, \\ t(x, u_1, \ldots, u_j) = t(x, u_1, \ldots, u_{j-1}) + \varphi(t(x, u_1, \ldots, u_{j-1}), u_j)dt. \end{cases} \tag{7.23}$$

We define the function D to be maximised:

$$D(x, u_1, \ldots, u_j) = \begin{cases} f_n(t(x, u_1, \ldots, u_j)) & \text{if } t(x, u_1, \ldots, u_j) \in K, \\ - \parallel t(x, u_1, \ldots, u_j) - c \parallel & \text{otherwise} \end{cases} \tag{7.24}$$

where c is the center of K. The first case corresponds to situations where the SVM function can be directly used to find an optimal control, and the second case to situations where a point is located inside a zone not defined by the learning set and the SVM function behaviour is not constrained (it is also used at the first iteration of the algorithm, when no SVM function is available).

The set of optimal controls $(u_1^*, \ldots, u_j^*)$ is the one solving:

$$\arg \max_{(u_1,\ldots,u_j)} D\left(x, u_1, \ldots, u_j\right). \tag{7.25}$$

For instance, we can use a simple gradient descent procedure to determine u^*. We build a series $u^{(k)}$, where u_i^k is the ith component of u^k, defined as follows:

$$u_i^{k+1} = u_i^k + \eta \frac{\partial D(x, u^k)}{\partial u_i}. \tag{7.26}$$

Parameter η fixes the magnitude of the update at each iteration.

7.4.2.3 Providing Fast and Compact Controllers

Using SVM to define boundaries of approximated viability kernels provides compact controllers because the function f is generally defined with a relatively small number of support vectors (a few percent of the points of the grid). Once the approximation is obtained, only the SVM function needs to be stored, that is to say the list of support vectors, their associated coefficient (α_i) and one additional parameter (b). Even if the size of the grid is huge, if the boundary is smooth enough, the number of support vectors will be small and easily storable.

SVM also enable the definition of fast controllers: the first reason is linked to the use of optimisation techniques described below, which prevent discretising the set of controls. In addition, we define in the next section an SVM based heavy controller that allows one to avoid the computation of the distance of a point to the approximation boundary, and only requires the evalution of the SVM function.

7.4.2.4 Simplified Algorithm

Algorithm (7.10) requires the computation of the distance from a point x to the set L_h^n. This computation is not straightforward and is time consuming (see Chap. 8.1 for more details). We noticed that, in the many cases we tested, the following rule to define K_h^{n+1} allows the limitation of the diffusive effect of the algorithm:

$$\begin{aligned} K_h^{n+1} = \{x_h \in K_h^n \text{ such that } & f_n(x_h + \varphi(x_h, u^*)dt) \geq -\delta \text{ and} \\ & (x_h + \varphi(x_h, u^*)dt) \in K\} \end{aligned} \tag{7.27}$$

where u^* is the control solving (7.25). We chose $\delta = 1$ because it limits the definition of the boundary to the -1 margin of the SVM function. This definition tends to satisfy a stricter condition than (7.12), but does not guarantee it.

7.4.2.5 Overall Algorithm

Algorithm 1 describes the overall simplified viability kernel approximation algorithm implemented on KAVIAR. Point x^* is the one obtained after using control u^* (or $(u_1^*, \dots, u_j^*)$ for a j time step optimisation).

Algorithm 1 Simplified viability kernel approximation SVM algorithm

```
𝒮 ← ∅
n = 0
K_h^1 ← ∅

{First iteration}
for all x_h ∈ K_h do
   if x_h^* ∈ K then
      𝒮 ← 𝒮 ⋃ (x_h, +1)
      K_h^1 ← K_h^1 ⋃ x_h
   else
      𝒮 ← 𝒮 ⋃ (x_h, −1)
   end if
end for

{Following iterations}
repeat
   n = n + 1
   Compute f_n(x) from 𝒮
   𝒮 ← ∅
   K_h^{n+1} ← ∅
   for all x_h ∈ K_h^n do
      if (f_n(x_h^*) ≥ −1) and (x_h^* ∈ K) then
         𝒮 ← 𝒮 ⋃ (x_h, +1)
         K_h^{n+1} ← K_h^{n+1} ⋃ x_h
      else
         𝒮 ← 𝒮 ⋃ (x_h, −1)
      end if
   end for
until K_h^{n+1} = K_h^n
Define L_h^n from f_n(x)
return L_h^n
```

7.4.2.6 SVM Viability Controllers

We use again the property that SVM functions can play the role of a barrier function close to the boundary of the approximation in order to define the SVM heavy viability controller algorithm:

Algorithm (one time step anticipation)

For given real values Δ and Δ_2, we define:

$$A_\Delta = \left\{x \in L_h^p \text{ such that } (f_p(x) \geq \Delta \quad \text{and} \quad d(x, K) > \Delta_2)\right\}. \tag{7.28}$$

Considering an initial state $x_0 \in A_\Delta$ and an initial control $u_0 \in U(x)$ chosen randomly, the procedure associates a control u_{n+1} at the $n + 1^{th}$ iteration as follows:

- If $(x_n + \varphi(x_n, u_n)dt) \in A_\Delta$, we keep the same control $u_{n+1} = u_n$
- Otherwise, $u_{n+1} = \arg\max_{u \in U(x_n)} f_p(x_n + \varphi(x_n, u)dt)$

For practical purposes, we can define more or less cautious controllers by anticipating on k time steps. Starting from a point x_n, we check for $i = 1, \ldots, k$, by applying k times control u_n, if the resulting point $t(x_n, u_n, \ldots, u_n)$ belongs to the set A_Δ. If yes, we move by one step with $u_{n+1} = u_n$. If not, we look for a sequence of controls that allows the point $t(x_n, u_{n+1}, \ldots u_{n+k})$ to remain inside A_Δ, and we apply control u_{n+1}. Then, the more the number of anticipation time steps k, the more cautious the controller. The previous algorithm then becomes:

Algorithm (cautious controller)

Considering an initial state $x_0 \in A_\Delta$ and an initial control $u_0 \in U(x_0)$ chosen randomly, the procedure associates a control u_{n+1} at the $n + 1^{th}$ iteration as follows:

- If $t(x_n, u_n, \ldots, u_n) \in A_\Delta$, we keep the same control $u_{n+1} = u_n$
- $(u_{n+1}, \ldots, u_{n+k}) = \arg\max_{u_{n+1},\ldots,u_{n+k} \in U(x_{n+1}),\ldots,U(x_{n+k})} f(t(x_n, u_{n+1}, \ldots u_{n+k})$ otherwise and we apply control u_{n+1}

There is no mathematical guarantee that this procedure allows the system to always remain inside K. However, it seems that anticipating on several time steps greatly increases the chances of staying inside K. The advantage of using an SVM function to define a controller is that it provides a fast and compact controller (see discussion above).

7.4.3 Resilience Basins Approximation and Resilient Policies of Action Definition Using SVM

Similarly to the viability kernel approximation, we can choose SVM as a particular classification function to approximate resilience basins (and hence resilience values). It enables one to use standard optimisation techniques to find an appropriate control u^*:

$$u^* = \arg\max_{u \in U(x)} f_n(x + \varphi(x, u)dt_{x,u}) \tag{7.29}$$

that leads to the "optimal" state x^*. We can also extend the procedure to several time steps, optimising at each iteration.

7.4.3.1 Simplified Algorithms

As in the viability kernel computation, we can use the SVM function f as a proxy for the distance to the boundary. This simplifies the computation, but again the threshold to accept or reject points must be tuned experimentally.

$$\begin{aligned}
&\textit{outer approx.}\\
&\quad H_h^{n+1} = \left\{x_h \in H_h \text{ such that } f_n(x_h^*) \geq -1 \text{ and } x_h^* \in K\right\}\\
&\textit{inner approx.}\\
&\quad H_h^{n+1} = \left\{x_h \in H_h \text{ such that } f_n(x_h^*) > +1 \text{ and } x_h^* \in K\right\}
\end{aligned} \tag{7.30}$$

For inner approximations, support vectors with label -1 are located inside the approximation and close to the SVM boundary. For outer ones, the approximation does not include support vectors of label $+1$.

7.4.3.2 Overall Algorithms

Algorithm 2 describes the resilience basins approximation algorithm with SVM. Details are given for an inner approximation, but an outer approximation algorithm can be easily built from it: instead of testing if $f_n(x_h^*) > +1$ (or $f_n(x_h) > +1$), we test if $f_n(x_h^*) \geq -1$ (or $f_n(x_h) \geq -1$). The algorithm returns the set of all the SVM functions f_q, $q = 1, \ldots, n+1$ that are used to compute the sets H_h^q belonging to successive resilience basins of decreasing resilience, representing the states that have a restoration cost lower of equal to $q.k.dc$, with k the number of time steps used at each iteration. The resilience values can be then easily computed as the inverse of the restoration costs.

7.4.3.3 SVM Optimal Controllers

We start with an initial point $x_0 \in C_h^q$. For each state x_i of the trajectory, we compute $n(i)$ the maximal value of n such that $x_i \in C_h^n$. If $n(i) > 0$, we define u_i^* as follows:

$$u_i^* = \arg \max_{u \in U(x_i)} f^{n(i)-1}(x_i + \varphi(x_i, u)dt_{x,u}). \tag{7.31}$$

If the approximation has been made using several time steps k, the algorithm must be slightly adapted. At the first iteration, we look for the number of time steps $j_0 \leq j$ needed to drive the point $t(x_0, u_1^*, \ldots, u_{j0}^*)$ to the next boundary $C_h^{n(i)-1}$:

$$\begin{aligned}
&\arg \max_{u_1,\ldots,u_{j_0} \in U(x)} f^{n(i)-1}(t(x_0, u_1, \ldots, u_{j_0})) \in C_h^{n(i)-1} \text{ and}\\
&\arg \max_{u_1,\ldots,u_{j_0-1} \in U(x)} f^{n(i)-1}(t(x_0, u_1, \ldots, u_{j_0-1})) \notin C_h^{n(i)-1}.
\end{aligned} \tag{7.32}$$

Then, we use (7.18), replacing $x_i + \varphi(x_i, u)dt_{x,u}$ by $t(x_i, u_1, \ldots, u_j)$.

Algorithm 2 Simplified resilience basins inner approximation SVM algorithm in the initial state space

$n = 0$

{Initialisation}
Get $\mathscr{S}$, $f_p(x)$, K_h^p and L_h^p using alg. (7.10)
$H_h^0 \leftarrow K_h^p$
$f_1(x) \leftarrow f_p(x)$

{Following iterations}
repeat
  $n \leftarrow n + 1$
  for all $x_h \in H_h$ **do**
    if $f_n(x_h) > 1$ **then**
      $\mathscr{S} \leftarrow \mathscr{S} \bigcup (x_h, +1)$
      $H_h^n \leftarrow H_h^n \bigcup x_h$
    else
      if $(f_n(x_h^*) > -1)$ and $(x_h^* \in K)$ **then**
        $\mathscr{S} \leftarrow \mathscr{S} \bigcup (x_h, +1)$
        $H_h^n \leftarrow H_h^n \bigcup x_h$
      else
        $\mathscr{S} \leftarrow \mathscr{S} \bigcup (x_h, -1)$
      end if
    end if
  end for
  Compute $f_{n+1}(x)$ from $\mathscr{S}$
  $\mathscr{S} \rightarrow \emptyset$
until $H_h^n = H_h^{n-1}$
return $\{f_q(x), q = 1, \ldots, n + 1\}$

Using an inner approximation of restoration cost to choose the controls guarantees that the procedure drives back the system to the viability kernel approximation. Once the system is viable again, we use the heavy controller procedure to maintain the system indefinitely in *Viab*(K).

7.5 Practical Examples Using KAVIAR

7.5.1 KAVIAR Software

KAVIAR is a research software under a GPL V3 license written in Java. Given the Euler approximation of a dynamical system, the application computes the corresponding viability kernel, capture basin or resilience basins approximation, using SVM. Once the approximation is obtained, the user can use an adequate controller (heavy or optimal). The software and a user guide are downloadable

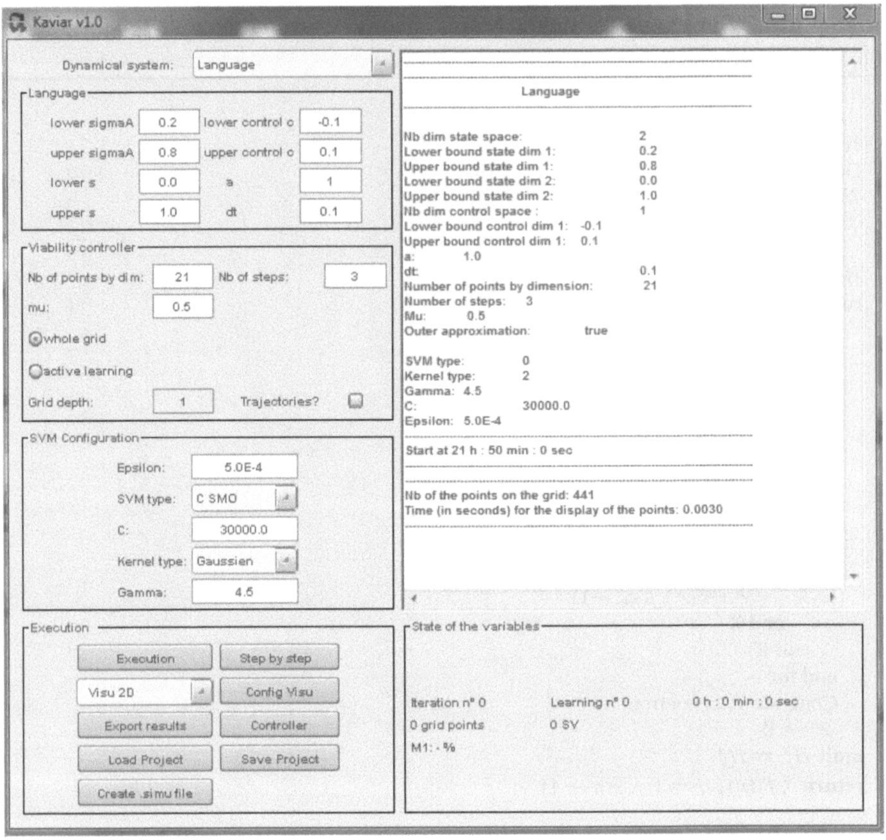

Fig. 7.4 KAVIAR graphical user interface

online.[3] KAVIAR works in two modes: a batch and a GUI mode. Figure 7.4 shows a snapshot of the graphical interface and Fig. 7.5 an example of a viability kernel with a heavy trajectory, and the associated controller panel. The batch mode allows faster computations, and can be used for a series of computations on machines with no graphical tools. For a quick start with the software, there exists a Java executable file which allows new users to make their first steps with KAVIAR. The source code is available in order to implement new models. In the user guide, one can find the prerequisites and the installation steps, how to run the program in batch and GUI modes, a description of the already implemented models and how to implement a new model.

[3]http://www.kaviar.prd.fr

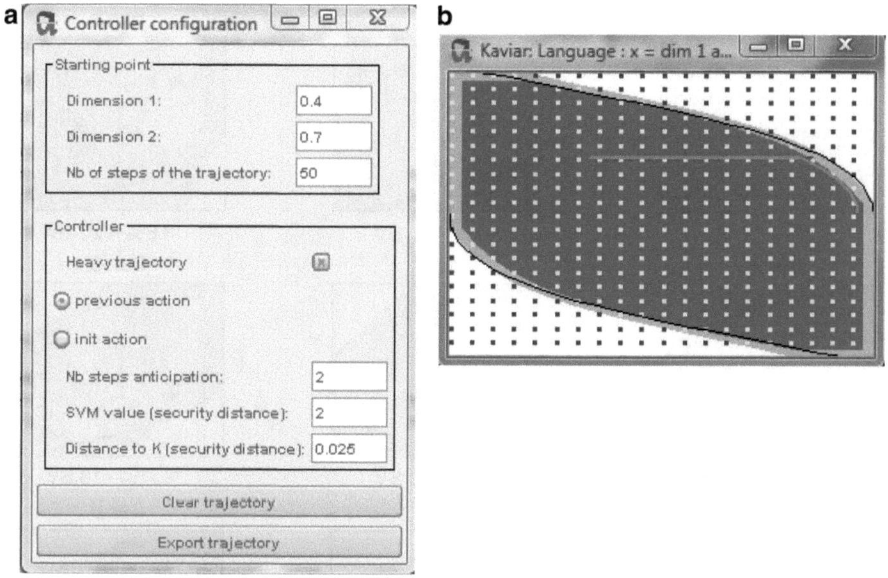

Fig. 7.5 (**a**) Controller panel, (**b**) visualisation frame

7.5.2 Example of Viability Kernel Approximations Using KAVIAR

7.5.2.1 Progressive Approximation

Figure 7.6 gives an example of a progressive approximation of the viability kernel, considering the Abrams and Strogatz model for language competition in two dimensions discussed in Chap. 3, with $a = 1$. Table 7.1 gives the parameters used and different statistics about the final approximation.

7.5.2.2 Finer Grid Gives a Better Approximation

For a given set of parameters (*dt* value and number of time steps), using a finer grid provides a better approximation. Indeed, the algorithm states that the approximation will converge toward the actual viability kernel when $h \to 0$. Figure 7.7 gives a comparison of two approximations, the first one using 19 points per dimension whereas the second one has 49 points per dimension.

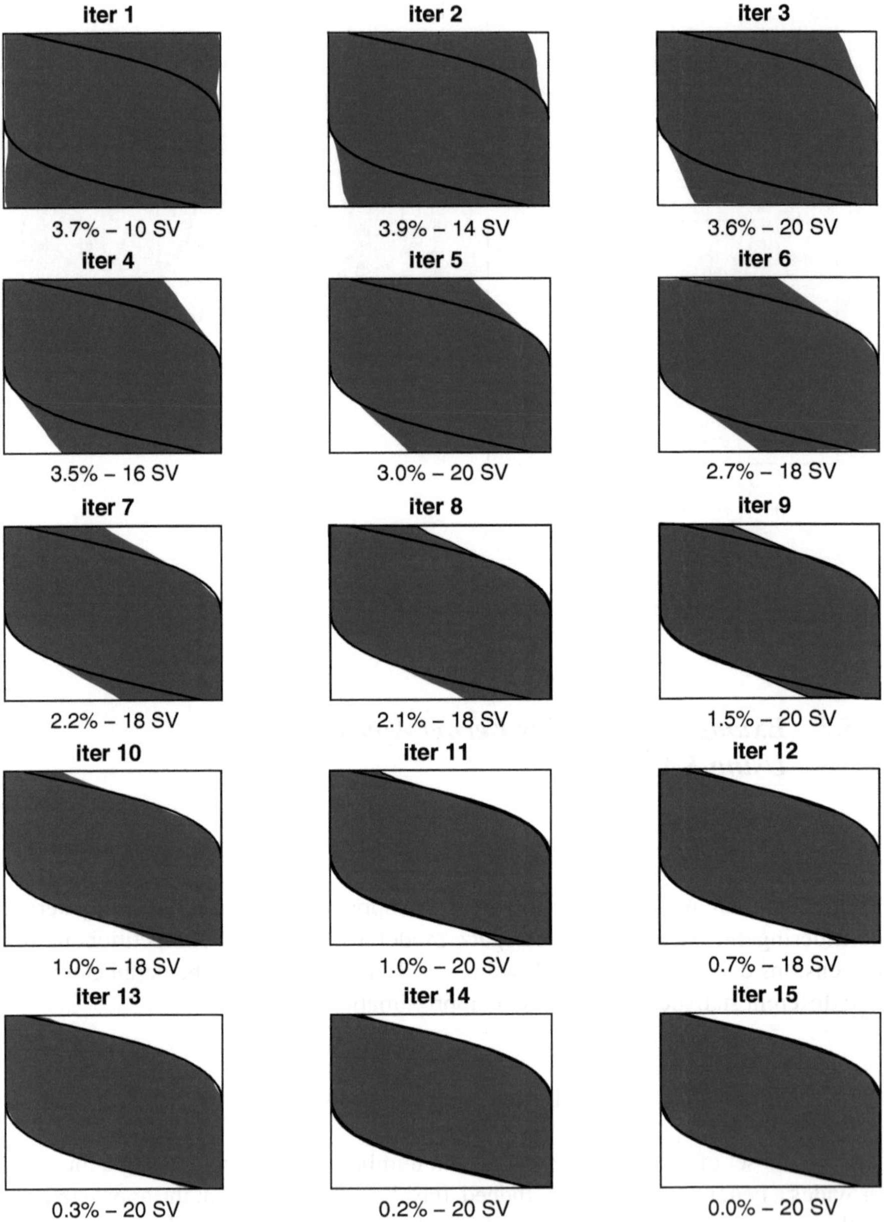

Fig. 7.6 Example of a progressive approximation of the viability kernel using KAVIAR. The different approximations L_n^h are represented in dark grey. The theoretical lines of the boundary of the viability kernel are represented in black. Below each graph, we indicate the number of support vectors needed to define the SVM function and the first figure in percentage represents the proportion of points of the grid (not represented here) that have had their label changed (from $+1$ to -1) between iteration $n-1$ and n

Table 7.1 Parameters and results of the simulations for the language problem

	Figure 7.6	Figure 7.7 (right)	Figure 7.8 (left)	Figure 7.9 (right)
nb points by dim	49	19	49	49
nb total of points	2401	361	2401	2401
nb of steps	6	6	1	1
dt	0.05	0.05	0.05	0.3
nb iterations	15	10	53	15
Final nb of SV	20	16	14	22
Time (s)	16	2	14	9

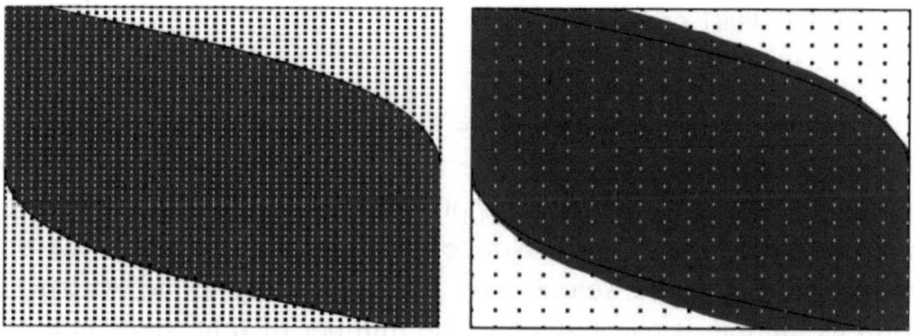

Fig. 7.7 Example of an approximation for two grid sizes (*left*) 49 points by dimension (*right*) 19 points by dimension. The points in white are the ones that are viable at the last iteration whilst the black ones are not viable

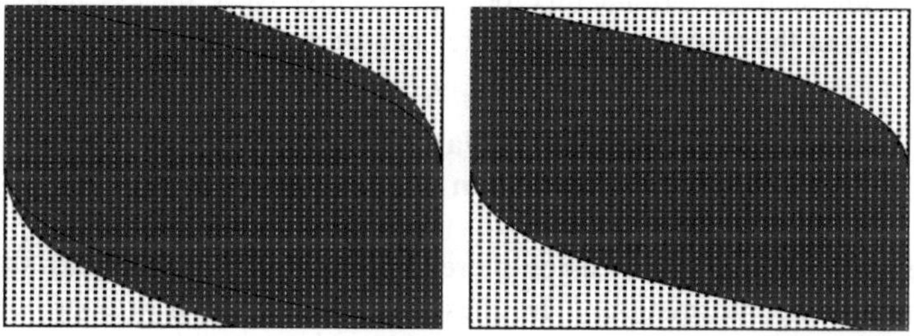

Fig. 7.8 Example of an approximation with two different time steps with the same $dt = 0.05$ (*left*) 1 time step (*right*) 6 time steps. The points in *white* are the ones that are viable at the last iteration whilst the black ones are not viable

7.5.2.3 More Time Steps Give a Better Approximation

For a given set of parameters (*dt* value and the number of points by dimension), using more time steps j provides a better approximation. Indeed, the algorithm states that the approximation will converge toward the actual viability kernel when $dt' \to 0$, with $dt' = j.dt$. Figure 7.8 gives a comparison of two approximations, the first one using 1 time step while the second one uses 6 time steps.

The number of time steps and the *dt* value must be chosen in a coordinated way. Figure 7.9 gives an example of two approximations with the same $dt' = 0.3$:

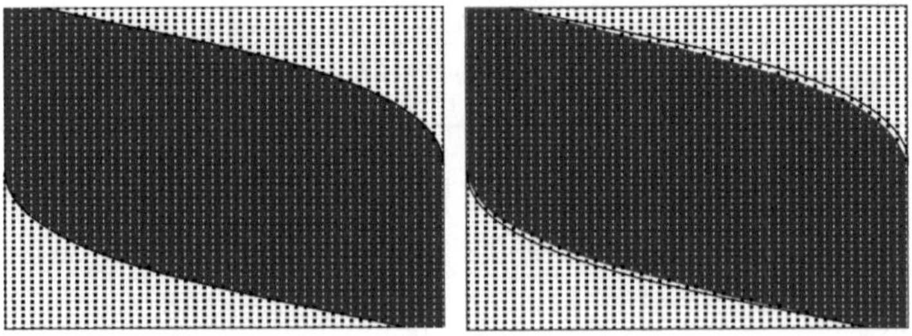

Fig. 7.9 Example of an approximation with two different time steps and *dt* values (*left*) $dt = 0.05$, 6 time steps (*right*) $dt = 0.3$, 1 time step. The points in white are the ones that are viable at the last iteration whilst the black ones are not viable

the first one considers $j = 1$ and $dt = 0.3$ while the second uses $j = 6$ and $dt = 0.05$. As explained previously, the first case provides a bad approximation as the continuous dynamics is poorly approximated by the discrete scheme, while the second case gives a good one. Notice that we cannot increase the number of time steps indefinitely: as the search for a viable control is done by using a simple optimisation technique, it can fail when the number of time steps is too large (in the experiments we performed with a simple gradient descent procedure, 8 steps was still a reasonable choice). Notice that there exists a viability kernel algorithm that uses a simulated annealing to perform a search for a viable control over a large number of time steps (see Bonneuil (2006) for details). Nevertheless, it degrades the performances of the algorithm as the search can be very time consuming.

7.5.2.4 Choosing an Appropriate *dt* Value Is Essential to Get a Better Approximation

We have seen previously that using more time steps and a finer grid allows one to obtain better approximations. The question that arises is: how does one choose "good values"? What really matters are the *dt* or dt' values relative to the grid size. We know that the finer the grid, the better the approximation. Then, according to the dimension of the problem (due to the dimensionality curse), we define the size of the grid: large when the dimension is small and rougher when the dimension increases. Once the size of the grid is defined, we have to choose a dt' value accordingly: if the value is too small regarding the size of the grid, the approximation will be much larger than the actual one; if it is too big, it can lead to smaller approximations. In the experiments we performed, we noticed that a dt' value that gives a point $x(t + dt')$ at a distance roughly equal to twice the maximal distance between two points of the grid of the initial point $x(t)$ seems to be a good choice.

One may wonder how to proceed when the distance from a point $x(t)$ and its successor $x(t + dt')$ is really different depending of the localisation of the state in the space. In that case, either we choose a value regarding the slowest dynamics or we can define an adaptive *dt* in function of the speed of the dynamics.

7.5.3 Computing Viable Policies of Actions with KAVIAR

One accesses the interface for computing trajectories by clicking on the button "Controller" in the KAVIAR interface. This interface offers the possibility to choose several parameters, in particular the caution margin Δ, and the number of anticipating steps k.

7.5.3.1 Choosing the Δ Value

The Δ value setting (defined formally in Sect. 7.4.2) depends on two aspects:

- How confident are we of the approximation? If the dt value, size of the grid and number of time steps have been carefully chosen, we know that the approximation is close to the actual viability kernel and a "small" Δ value should be enough to keep the trajectory inside K indefinitely
- How fine is the grid? The SVM function value $f_p(x)$ of a point depends on the number of points used to train the SVM. The finer the grid, the slower the SVM value decreases by going away from the boundary. However, there is no general rule that allows one to link the SVM function value with the distance to the boundary. The only way to get an idea of how large Δ is, is to either look at it when the system is in low dimension, or to compute the number of points situated in the approximation but not in A_Δ in larger dimension

Figure 7.10 presents an example of a trajectory starting from the approximation of Fig. 7.6, with an initial point $x_0 = (0.25, 0.95)$, during 300 time steps, by anticipating on $k = 1$ time step, choosing $\Delta = 3$ or $\Delta = 30$. For the second case, we can see that the initial point is outside A_Δ: from the beginning, we look for controls that will allow the trajectory to reach a safer area. In the two cases, as the approximation is close enough to the actual boundary, the trajectory remains inside K: the largest Δ value makes the trajectory go much further away from the boundary.

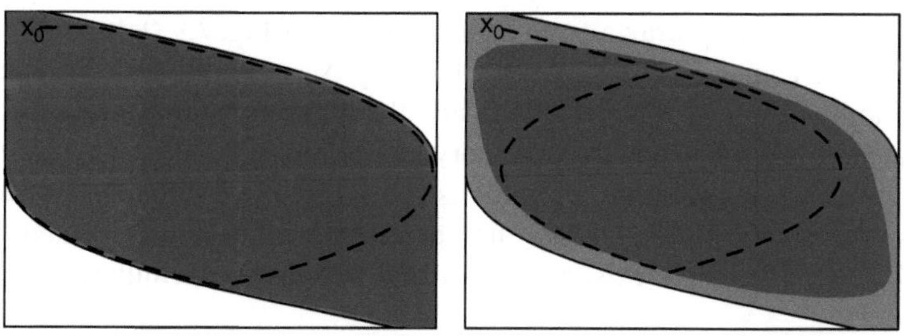

Fig. 7.10 Example of trajectories on 300 time steps, with (*left*) $\Delta = 3$ (*right*) $\Delta = 30$

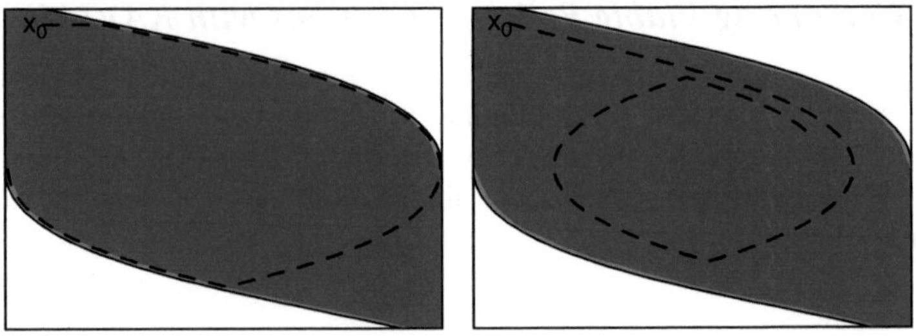

Fig. 7.11 Example of trajectories on 300 time steps, anticipating on (*left*) $k = 1$ time step (*right*) $k = 10$ time steps

7.5.3.2 Tuning the Number of Anticipating Time Steps k

Increasing the number of anticipating time steps k allows the definition of more cautious controllers, by preventing the trajectory from getting too close to the approximation boundary. Figure 7.11 compares two different examples of trajectory (keeping the same parameters as above, with $\Delta = 3$): one anticipating on $k = 1$ time step and the other on $k = 10$ time steps. As for the Δ value, a large k value produces a trajectory that stays away from the boundary.

7.5.3.3 Using the Controller to Check the Approximation

The SVM heavy controller algorithm can be used to "check" the viability kernel approximation. By definition, inside the viability kernel, we know that there exists at least one trajectory that will maintain the system inside K, and that there are none outside the kernel. Then, once we get an approximation, we can draw some trajectories from points located inside the approximation: if they stay inside the approximation, at least for those initial points and those of the trajectory, the approximation performs well. On the contrary, if they leave K after some time, there is probably something wrong with the approximation. Figure 7.12 gives an example of a trajectory drawn from approximation of Fig. 7.9 (left panel) which is larger than the actual viability kernel. Choosing $\Delta = 10$, the trajectory leaves the actual viability kernel whilst staying inside the approximation. When the controls are updated, it is too late and the trajectory leaves K.

In the same way, when the dt value is too large, the continuous dynamics are poorly approximated by the discrete ones, even if the initial state is located inside the true viability kernel, there exist no actions that will drive indefinitely the trajectory inside K.

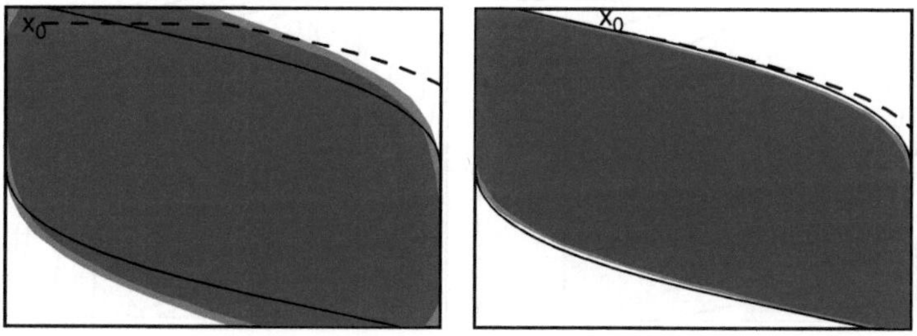

Fig. 7.12 Example of trajectories that allow one to see that the actual viability kernel is poorly approximated: (*left*) when the approximation is too large, (*right*) when the time step *dt* is too large

7.5.4 Examples of Approximated Resilience Basins with KAVIAR

7.5.4.1 Using Different Cost Functions

We illustrate the software again on the Abrams and Strogatz model, with parameter $a = 1$. We consider the following cost function that associates a minimal cost to a state x and a control u:

$$\ell(x, u) = \gamma(x, u) + g(x) \tag{7.33}$$

where

$$\gamma(x, u) = \min_{x(.)} \left(c_1 \int \chi_K x(\tau) d\tau \right) \tag{7.34}$$

with $\chi_K x(\tau) = d(x(\tau), K)$ when $x(\tau) \notin K$ and 0 otherwise, and

$$g(x) = \min_{x(.)} \left(c_2 \int \chi_V x(\tau) d\tau \right) \tag{7.35}$$

with $\chi_V x(\tau) = 1$ when $x(\tau) \notin Viab(K)$ and 0 otherwise (we take the approximation of $Viab(K)$ here).

KAVIAR first computes the viability kernel associated with the system and then, the resilience basins (either inner or outer approximations). Figure 7.13 gives an example of inner approximation of resilience basins and Fig. 7.14 an example of outer approximations, considering two sets of parameters: $c_1 = 0$ and $c_2 = 1$, which correspond to a cost function where we only take into account the time the system is outside the viability kernel, and $c_1 = 20$ and $c_2 = 1$, which add a part that corresponds to the distance to the viability constraint set. For each figure, one level curve corresponds to a cost variation of 4.8 (and hence to a resilience variation after perturbation of $\frac{1}{4.8} \approx 0.2$). Table 7.2 summarises the parameters used to compute approximations and gives some results.

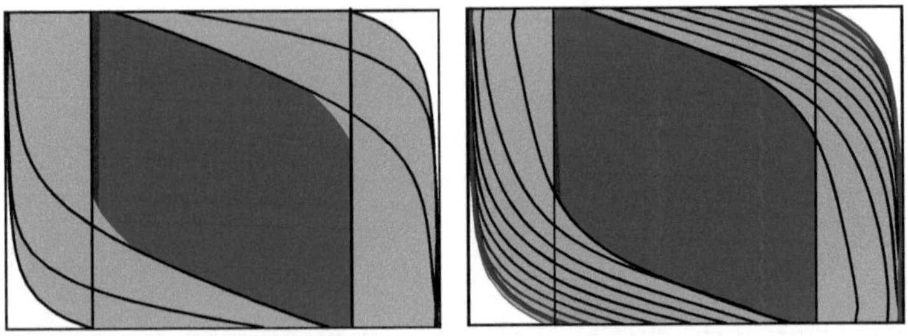

Fig. 7.13 Example of inner approximation of resilience basins with (*left*) $c_1 = 0$ and $c_2 = 1$ (*right*) $c_1 = 20$ and $c_2 = 1$. The viability kernel is represented in *dark grey*. Level *lines* represent the limits of successive resilience basins, and the states located between the level curves (*light grey area*) are resilient. States located behind the last level curve are not resilient

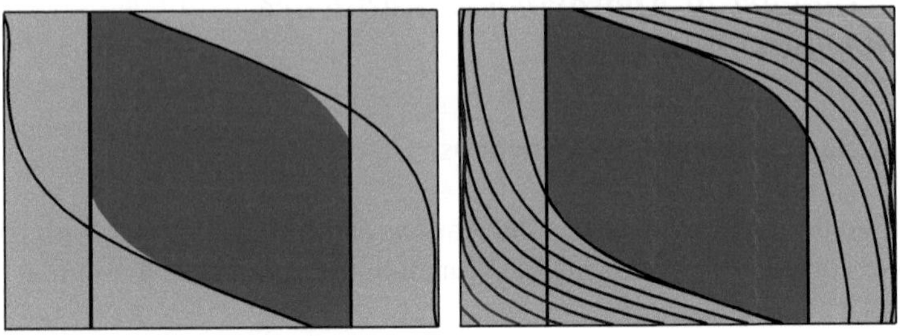

Fig. 7.14 Example of outer approximation of resilience basins with (*left*) $c_1 = 0$ and $c_2 = 1$ (*right*) $c_1 = 20$ and $c_2 = 1$

Table 7.2 Parameters and results of the cost approximation for the language problem 3

	Figure 7.13 (left)	Figure 7.13 (right)	Figure 7.14 (left)	Figure 7.14 (right)
nb points by dim	81	81	81	81
nb total of points	6561	6561	6561	6561
nb of steps	6	6	6	6
dt	0.05	0.05	0.05	0.05
dc	0.075	0.3	0.075	0.3
Maximal cost	14.4	62.4	11.7	59.4
nb of level lines	31	34	26	33
nb iterations	51	54	46	53
Time (s)	483	360	341	294

7.5.4.2 Choosing Appropriate Parameter Values Is Essential to Get Better Resilience Basins Approximations

As for the viability kernel approximation, the following parameter values are essential to obtain "good" approximations:

- Size of the grid: a finer grid gives better approximations

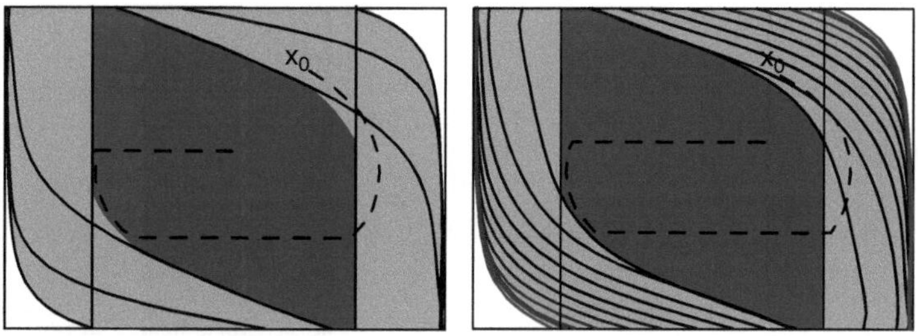

Fig. 7.15 Example of optimal trajectories (in *dotted line*) with (*left*) $c_1 = 0$ and $c_2 = 1$ (*right*) $c_1 = 20$ and $c_2 = 1$

- Number of time steps: an appropriate number of time steps gives better dynamic approximation while computing accurate viable / resilient controls
- *dt* values, chosen accordingly to the number of time steps and the size of the grid

For the resilient basins approximation, one has to choose carefully the *dc* value as well: this parameter is closely linked to the *dt* value and the same comments apply. Once the size of the grid is set, the *dc* value must be defined such that its value is neither too small (in that case, the resilient set will be smaller than the actual one), nor too big (the dynamics will be poorly approximated, which can lead to incorrect results). We noticed that a *dc* which gives a successor at a distance equal to 1–2 times the maximal distance between 2 points of the grid seems to be a good choice.

7.5.5 Computing Resilient Actions with KAVIAR

7.5.5.1 Optimal Controllers for Different Cost Values

Figure 7.15 gives an example of optimal trajectories (when the trajectory is outside the viability kernel approximation) starting from an initial resilient state x_0, and having a heavy trajectory during 750 time steps, for the different cost functions defined in the previous section. Note that, depending of the cost functions considered, the trajectories are slightly different: trajectory (a) minimises the time of restoration while trajectory (b) minimises a more complex cost.

7.5.5.2 Controllers as a Way to Check the Approximations

In the same way as for the heavy viable controller, the optimal controller procedure can be used as a way to "check" if the approximation is accurate. For instance, using the inner approximation algorithm, we know that the optimal trajectory will reach the viability kernel approximation. So, considering several initial points and checking if we can define a series of controls that drives the system back to

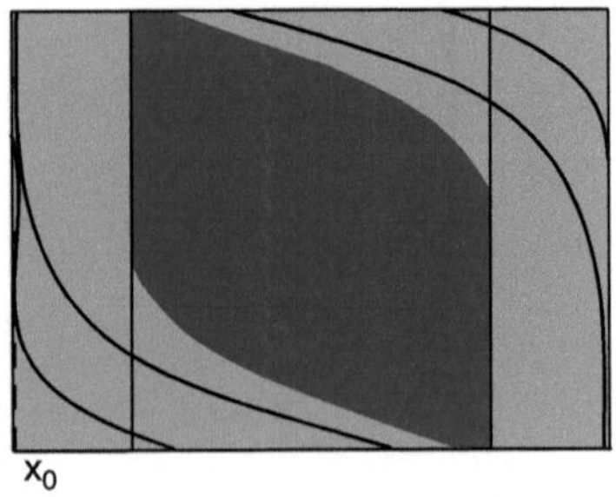

Fig. 7.16 Example of a trajectory starting from an outer approximation which fails to come back to the viability kernel

the kernel is a way to gain confidence on the approximation: if all the resulting trajecories are back, the approximation looks good enough: if not, there is probably something wrong with the approximation (and the chosen parameters must be more carefully set).

Figure 7.16 presents an example of a trajectory starting from a non-resilient state, but located inside the approximated resilient set. Results are those obtained in Fig. 7.14 (outer approximation). The initial state $x_0 = (0, 0)$ is obviously not resilient (details about the dynamics can be found in Chap. 3), but located inside the approximated resilient state. The algorithm then tries to drive the state back to the viability kernel, without success, and stops when the current cost of restoration is too far away from the expected one.

7.6 Discussion: Limits and Future Research

This chapter aimed at completing the user guide, by providing more detailed information about the algorithms, and also some practical clues to obtain "good" approximations and design "good" controllers. We illustrated the algorithm and software on the Abrams and Strogatz model as an example, taking different parameter values and explaining the results. We showed that, when parameters are carefully chosen, accurate and reliable approximations are obtained.

Algorithms implemented on the software allow users to deal with systems in high control dimension space, as they are no longer discretized. Nevertheless, the limitations on the state space still remain and restrict practically the approach to low dimension space (5 or 6 dimensions at most). Using SVM as a classification function provides a compact formulation of the approximated viability kernels and resilience basins, and then allows the definition of fast controllers: even if the approximation step is time and space consuming, the final approximations can be stored cheaply and provide quickly a viable or optimal trajectory.

Two intrinsic limitations are related to the choice of SVM as a classification procedure:

- The most important one is that we cannot guarantee that the result of the SVM procedure satisfies the conditions of the theorem. Indeed, depending on the geometry of the problem, and on the parameter of the SVM solving algorithm, some artificial parts of boundary might be created. Loosli et al. (2008) studied this phenomenon and derived an algorithm that modifies automatically the SVM parameters to avoid this situation. Unfortunately, although it provides good results in practice, it gives no formal guarantee either. Hence it might be preferred to tune the SVM parameters by hand, in order to keep their properties right.
- Another important drawback is the difficulty to compute the distance from a point to the boundary defined by the SVM (see Chap. 8.1 for a discussion). In the algorithm, we use the SVM function as a proxy for this distance, but this is not entirely satisfactory. Indeed, this proxy can be of very poor quality, especially when the point is far from the boundary.

Therefore, a lot of research is still open to find learning procedures which would be more adequate to the problem, and keeping the promises of parsimony that made us choose SVM initially. More generally, we can think of other important research directions:

- Introducing heterogeneous grids, which are more refined close to the boundaries. This will probably allow us to gain some dimensions but may not be enough to break the curse. Anyway, the price for higher dimensions is probably a dramatic decrease of the accuracy of the approximations.
- Adapting the algorithm for stochastic systems: our initial assumption was that the systems of interest are deterministic. The next step is to design algorithms that can take into account noise on the dynamics.

Finally, it must be underlined that KAVIAR is research prototype software. Some of the features are not optimal: for instance, the graphical user interface can be improved, the execution can be slow and there may be some bugs left. We hope that it will evolve and improve over time, with the input of more and more users.

References

Aubin JP (1991) ViabilityTheory. Birkhäuser, Basel

Aubin JP (2001) Viability kernels and capture basins of sets under differential inclusions. SIAM J Contr Optim 40(3):853–881

Aubin JP, Frankowska H (1996) The Viability Kernel Algorithm for Computing Value Functions of Infinite Horizon Optimal Control Problems J Math Anal Appl 201(2):555–576

Bellman R (1961) Adaptive Control Processes: A Guided Tour. Princeton University Press, Princeton

Bokanowski O, Martin S, Munos R, Zidani H (2006) An anti-diffusive scheme for viability problems. Appl Numer Math 56(9): 1147–1162

Bonneuil N (2006) Computing the viability kernel in large state dimension. J Math Anal Appl 323:1444–1454

Boser BE, Guyon I, Vapnik V (1992) A training algorithm for optimal margin classifiers. In: Proceedings of the Fifth Annual Workshop on Computational Learning Theory, Pittsburgh, USA, pp 144–152

Chapel L, Deffuant G (2007) SVM viability controller active learning: application to bike control. In: Proceedings of the IEEE International Symposium on Approximate Dynamic Programming and Reinforcement Learning (ADPRL'07), Honolulu, USA, April 1–5, pp 193–200

Chapel L, Deffuant G (2011) Inner and Outer Capture Basin Approximation with Support Vector Machines. To appear in: Proceedings of the 8th International Conference on Informatics in Control, Automation and Robotics, Noordwikerhout, The Netherlands

Cristianini N, Shawe-Taylor J (2000) Support Vector Machines and other kernel-based learning methods. Cambridge University Press, Cambridge

Deffuant G, Chapel L, Martin S (2007) Approximating viability kernels with support vector machines. IEEE Trans Automat Contr 52(5):933–937

Loosli G, Deffuant G, Canu S (2008) BALK : bandwidth autosetting for SVM with local kernels. Application to data on incomplete grids. Conférence d'apprentissage

Martin S (2004) The cost of restoration as a way of defining resilience: a viability approach applied to a model of lake eutrophication. Ecol Soc 9(2)8

Platt JC (1999) Fast training of support vector machines using sequential minimal optimization. In: Schölkopf B, Burges CJC, Smola A (eds) Advances in kernel methods – support vector learning. MIT, Cambridge, pp 185–208

Saint-Pierre P (1994) Approximation of the viability kernel. Appl Math Optim 29:187–209

Scholkopf B, Smola A (2002) Learning with Kernels: support vector machines, regularization, optimization, and beyond. MIT, Cambridge, MA, USA

Vapnik V (1995) The nature of statistical learning theory. Springer, New York

Chapter 8
Geometric Robustness of Viability Kernels and Resilience Basins

Isabelle Alvarez and Sophie Martin

8.1 Introduction

The definition of resilience described in Chap. 2 applies to dynamical systems whose dynamics are modelled by controlled differential equations and in which some properties of interest are defined by a subset of the state space (the constraint set). These dynamical systems, when they model environmental or socioeconomic systems, are subject to uncertainties. Moreover, the properties of interest are rarely known with absolute certainty and accuracy. Viability theory can take into account only a part of these uncertainties, considering a set of velocity vectors rather than a single vector. Therefore, performing sensitivity analysis (like in Saltelli et al. 2000) appears as a good solution to assess the impact of the other parameters of the dynamics, and also the impact of slight modifications of the boundary of the constraint set on the viability kernel and its capture basin.

As seen in Chap. 2, the resilience value is infinite inside the viability kernel, but outside this set it can switch to a finite value or even to zero. If a perturbation leads the system to a state of zero resilience value, there is no hope of driving it back eventually to the viability kernel.

This is the reason why an explicit study of the robustness of states and trajectories is so important in viability studies. In this chapter, we propose a geometric method to appraise the robustness of the results given by a viability study. Intuitively, the robustness or the risk associated with a given state depends on its position relative to the boundary of the viability kernel or the resilience basin.

I. Alvarez (✉)
Cemagref - LISC, 24 av. des Landais 63172 Aubière, France

LIP6, UPMC, 5 place Jussieu 75005 Paris, France
e-mail: isabelle.alvarez@cemagref.fr

S. Martin
Cemagref - LISC, 24 av. des Landais 63172 Aubière, France
e-mail: sophie.martin@cemagref.fr

G. Deffuant and N. Gilbert (eds.), *Viability and Resilience of Complex Systems*, Understanding Complex Systems, DOI 10.1007/978-3-642-20423-4_8,

These geometric concepts and their interest according to the resilience issue are described in Sect. 8.2, with a simple example. Prerequisite, correct use of the method and algorithms are described in Sect. 8.3. The geometric study is illustrated on the PATRES case study of language competition in Sect. 8.4 (see Chap. 3). It highlights the role the robustness information can play when controlling the system.

8.2 Geometric Criteria of Robustness in Viability and Resilience Analyses

8.2.1 Geometric Description of Relevant Sets

As in Chap. 2, we consider a dynamical system and we suppose that a property of interest of this system is defined as a constraint set (subset of the system's state space). The viability kernel is formed by all the states belonging to the constraint set that are viable, that is, the subset of initial points from which it is possible to maintain the system inside the constraint set. The resilience basin is defined as the capture basin of the viability kernel, that is the set of all states from which it is possible to reach the viability kernel in finite time.

The volume of these sets is useful to qualify the robustness of the system. The smaller they are, the less robust the system is. For example, if the volume of the viability kernel is very small compared with the volume of the constraint set, then it will be difficult to maintain the system in this desired set. If the volume of the resilience basin is very small compared with the volume of the state space (when its size is finite), then the system itself is not very resilient.

The variation of these sets with small modifications of the parameters is also a valuable piece of information. In particular, if small modifications of the constraint set lead to catastrophic modifications of the viability kernel (empty or very small set), then the viability study is not robust to uncertainties in parameters.

Besides this sensitivity analysis of the model, the geometric description provides more information concerning the robustness to uncertainty or measurement error of the state and control variables. Figure 8.1a shows two viability kernels with the same volume. Obviously a dynamical system that evolves in the top left kernel is less resistant to perturbation than a dynamical system that evolves in the top right kernel.

A useful indicator of the shape of the viability kernel or of the resilience basin is the diameter of the largest maximal ball. Maximal balls inside a set are open balls that are not contained in any larger ball inside the set. Centres of maximal balls form the skeleton (see Serra 1988, for more information about mathematical morphology concepts). The largest maximal ball is the largest ball inscribed in the set. The centre of the largest maximal ball is the farthest point from the boundary of the set. With Euclidean distance, the largest maximal ball is a sphere. When the centre of the

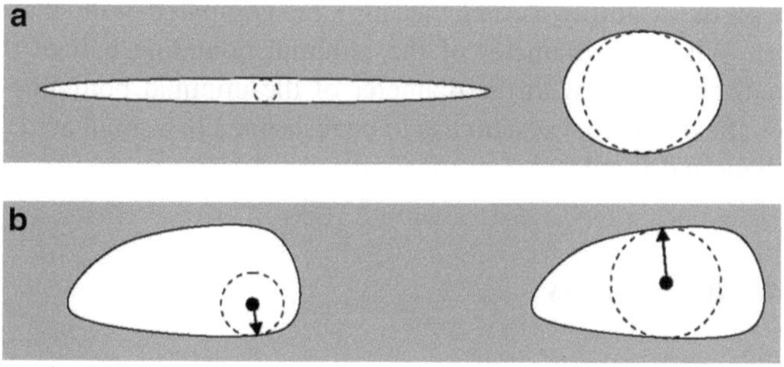

Fig. 8.1 Two examples of situations where the geometric description brings useful information. (**a**) Two kernels with the same volume. A system evolving in the left kernel cannot resist even small disturbance along the vertical axis. *Dot circles* show the respective largest maximal balls of both sets. (**b**) Two points in the same kernel. The *dashed circle* shows the largest perturbation that keep the point inside the viability kernel

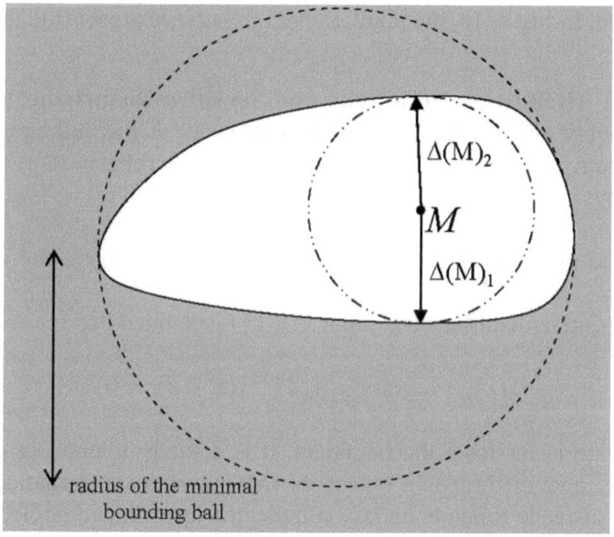

Fig. 8.2 Geometric indicators for the viability kernel or the resilience basin: the radius of the kernel minimal bounding ball and the radius of the largest maximal ball inside the kernel. M is the centre of the maximal ball, and $\Delta(M)_1$ and $\Delta(M)_2$ are the sensitive disturbances at point M

largest maximal ball is close to the boundary of the set, then every point is close to the boundary, as is the case in Fig. 8.1a on the left.

The diameter of the largest maximal ball can be compared with two base characteristics of the viability study, as shown in Fig. 8.2. When the diameter of the largest maximal ball is small compared with the diameter of the minimal bounding ball of the set, the system is very sensitive to even small disturbances. In the case of a viability kernel or a resilience basin, this means that the system is not robust to the uncertainties in the state variables.

The size of the viability kernel can also be compared with the size of the constraint set. When the diameter of the minimal bounding ball of the viability kernel is small compared with the diameter of the minimal bounding ball of the constraint set, the dynamical system has to be restrained in a small area. This makes control much more difficult.

8.2.2 *State Robustness*

When a system starts from a state inside the viability kernel, it is possible to maintain its evolution inside the constraint set with certainty, whereas if the system starts inside the resilience basin, it is possible to reach the viability kernel in finite time. These are the properties that characterize the viability kernel and the resilience basin. Nevertheless all states are not equal in regard to other respects: state utility, control cost, but also robustness to uncertainty or measurement error and robustness to perturbation. In Fig. 8.1b, the point on the left is far less robust than the point on the right.

Definition 8.1 (Robustness of a State and Sensitive Disturbance). Let K be a viability kernel or a resilience basin in the state space E (or in the extended phase space with control variables). Let x be a state in K. The robustness $s(x)$ of the state x is defined by:

$$s(x) = \max\{\alpha \geq 0;\ \forall y \in E,\ d(x, y) < \alpha \Rightarrow y \in K\}.$$

A minimal sensitive disturbance at point x, $\Delta(x)$ is defined by:

$$\Delta(x) = argmin_{\delta \in E}\{\|\delta\|\ ,\ s(x + \delta) = 0\}.$$

When a state is far from the boundary, it is actually tolerant to error in state determination, and in the same extent its robustness to perturbation is high. In Fig. 8.1b, the distance to the boundary is indicated by a dashed circle. It is obvious that the robustness value of the right state is greater than the robustness value of the left state. When a small perturbation is applied to the left point, it leaves the viability kernel. The point on the right can resist larger disturbances. Inside the viability kernel, where states are undifferentiated as to resilience, the concept of robustness is more appropriate to describe the situation, since the robustness of a state is the size of the largest perturbation that keeps the system in the viability kernel.

8.2.3 *Trajectory Robustness*

The distance map is used to define the robustness of a state, but it can also be used to propose a family of robustness definitions on trajectories. Since the distance map gives the distance to the boundary at each point, it is possible to use this information to define robustness indicators at the level of a trajectory. Several definitions can be proposed, depending on the risk perception of the manager.

Definition 8.2 (Geometric Robustness of a Trajectory). Let $x : t \longmapsto x(t)$ be a trajectory in the viability kernel or resilience basin K. We note Γ_K its boundary. Let f be a function from the set of continuous real-valued functions to $\mathbb{R}$. The f-geometric robustness value of u on the time interval $[0, T]$ is:

$$r_f(x) = f\left(\{t \in [0,T] \mapsto d(x(t), \Gamma_K)\}\right).$$

For example, the most risk-adverse indicator is the minimum of the robustness on the trajectory, with $f = \min$. But it has some drawbacks, since it does not take into account the time during which the robustness is low. So the function $f = \text{mean}$ can also be used as an average value to consider the robustness of a trajectory. (This definition has its own drawbacks, since a trajectory that leaves the set has a non-zero robustness).

Definition 8.3 (Min-robustness and Mean-robustness of a Trajectory). Let $x : t \longmapsto x(t)$ be a trajectory in the viability kernel or resilience basin K. We note Γ_K its boundary. The min-robustness value of x on the time interval T is:

$$r_-(x) = \min_{t \in [0,T]} \{d(x(t), \Gamma_K)\}$$

The mean-robustness value of x on the time interval $[0, T]$ is:

$$r_m(x) = \frac{1}{T} \int_{t \in [0,T]} d(x(t), \Gamma_K) dt.$$

Other definitions can be proposed with a discounting rate for future robustness values, in a similar way to what is done for delayed rewards or payments (which are worth less than if they were paid at present time). For example, the control policy can take into account the fact that it is possible to modify a trajectory in the future. Therefore the robustness at the present time of a state reached in the future is less critical than the robustness of the present state. (The robustness value of a future state is considered to be less than the robustness value of the present state with a positive discount factor).

Definition 8.4 (Discounting Robustness). Let α be a discount factor. The discounting robustness value of x on the time interval $[0, T]$ is:

$$r_\alpha(x) = \frac{1}{T} \int_0^T d(x(t), \Gamma_K) e^{-\alpha t} dt.$$

With these definitions, both trajectories x and y starting from A in Fig. 8.3 have a strictly positive robustness value for all time intervals, and the robustness of x

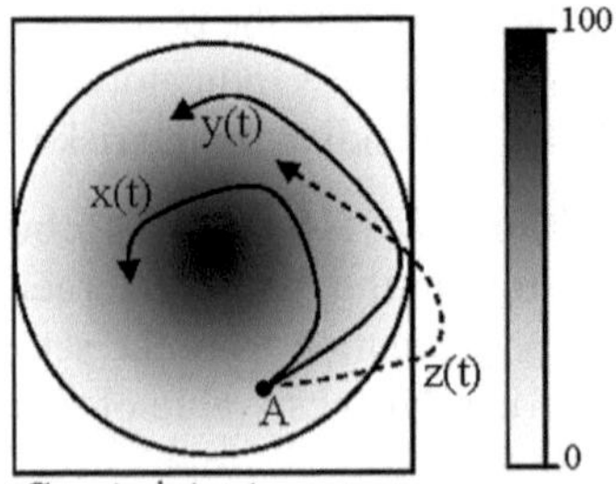

Fig. 8.3 Geometric robustness. Trajectory $x(t)$ is more robust than trajectory $y(t)$. Trajectory $z(t)$ has a min-robustness value of 0, since it leaves the viability kernel

is greater than the robustness of y, for the min-robustness, the mean-robustness or the discounting robustness (actually all f-robustness with f monotone). The robustness of trajectory z is defined on $[0, T_1]$, where T_1 is the time it exits the viability kernel. Its min-robustness value on the time interval $[0, T_1]$ is zero.

All these definitions can be adapted to the discrete case. For example, the discounting robustness in the discrete case is:

$$r_\alpha(x) = \frac{1}{T}\sum_{t=0}^{T} d(x(t), \Gamma_K)\frac{1}{(1+\alpha)^t}.$$

Trajectory robustness can be used to compare different evolutions of the system, starting from the same state point at some time t_0. For example, in Fig. 8.3, a manager will disregard trajectory z, because it leaves the viability kernel. Since trajectory x dominates y for all robustness indicators, the manager should follow the sequence of control of trajectory x.

8.2.4 A Simple Example

The geometric approach can be better understood when it is applied to a simple example, like the problem of lake eutrophication (see Carpenter et al. 1999, for an extensive description).

Clear water or water in the oligotrophic state provides ecosystem services such as freshwater, irrigation supplies, etc. of much higher economic value than turbid water or water in the eutrophic state. But many lakes have experienced sudden shifts from oligotrophic to eutrophic states. Phosphorus is the most critical nutrient for the eutrophication of lakes. Excess phosphorus is imported by farms in the form of fertiliser and animal feed supplements. Most of the phosphorus accumulates in soil, and may then be transported to streams and lakes during runoff events associated with snow melt or rainstorms. Knowing the dynamics of lake eutrophication, the available regulatory laws, the present concentration of phosphorus in the lake and the present amount of inputs, the issue is to determine whether the lake can remain in an oligotrophic state or if it is doomed to become eutrophic in finite time. Simple models describe the lake time evolution with few variables: L, the amount of phosphorus inputs and P the phosphorus concentration in the lake.

Agriculture requires a minimum value for L, whereas oligotrophic state requires a maximum value for P. In this model, regulatory laws are constraints on $\frac{dL}{dt}$: this means that the law cannot set the maximum amount of phosphorus inputs, but rather imposes a decrease of the phosphorus inputs, for example by a percentage each year, with a maximum allowed.

With these simple assumptions, it is possible to compute the viability kernel for the lake eutrophication problem. The viability kernel (see Fig. 8.4) is the subset of the (L, P)-plane that gathers all states (L, P) such that there exists at least one regulatory law that allows the oligotrophic state to be maintained (Martin 2004).

The information about the distance to the viability kernel boundary is valuable because of measure uncertainties or exogenous disturbances (which, for instance, cause a sudden increase in phosphorus concentration). Figure 8.4 shows the viability kernel with the level curves of the distance to its boundary. The geometric study shows in what cases a state is dangerously close to the decision boundary. For example, point B stands very close to the boundary compare to A: it is less robust to perturbation.

The projection onto the boundary shows the direction of the most dangerous disturbance. Figure 8.4 shows that when the lake is in state B, a small increase of phosphorus input (L) combined with a very small change in the phosphorus concentration (P) can shift the state of the lake outside the viability kernel. This means that in a case of such a small disturbance, at some time in the future, whatever controls are applied, the lake will experience eutrophication.

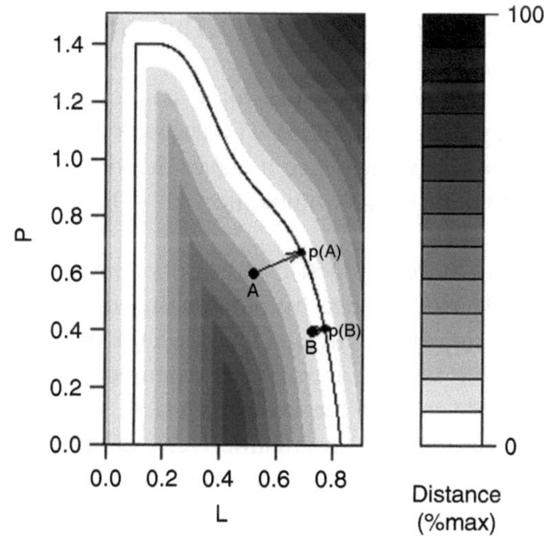

Fig. 8.4 Distance map for the viability kernel boundary of the lake eutrophication problem, with the following constraints: $0 \leq P < 1.4$, $0.1 < L < 1$, $\left|\frac{dL}{dt}\right| \leq 0.09$. The *black line* is the viability kernel boundary. $p(A)$ and $p(B)$ show the direction and size of the minimal sensitive disturbance at point A and B, respectively

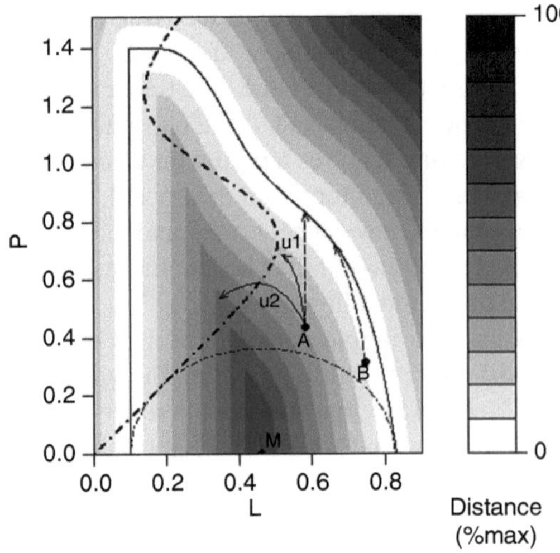

Fig. 8.5 Trajectories for the lake eutrophication problem, corresponding to constant negative or null control. The *dot dashed line* is the set of equilibriums. Long dashed trajectories that leave the viability kernel are not robust. Point M is the centre of the largest maximal ball

The geometric study also gives global information concerning the problem. In Fig. 8.5, point M is the centre of the largest maximal ball of the viability kernel. It is the farthest point from the boundary with $d(M, \Gamma) \approx 0.36$. The corresponding diameter $0.36 \times 2 = 0.72$ can be compared with the size of the constraint set (0.9 along L). Because this viability kernel has a reasonable size, it should be easy to find an action policy that guarantees an oligotrophic state.

The geometric study gives information about trajectories. We consider controls of the type $\frac{dL}{dt} = -\alpha L(0)$, with $\alpha \geq 0$, $L(0)$ being the amount of phosphorus inputs at the initial time. Both trajectories u_1 ($\alpha = -5\%$) and u_2 ($\alpha = -15\%$) starting from A with a constant (negative) control have a strictly positive value, either for the min or the mean-robustness. Both min and mean-robustness values of u_2 are greater than the ones of u_1. The vertical trajectory coming from A is not robust for the min-robustness. It corresponds to the trajectory with no phosphorus input variation ($\alpha = 0$). If the initial state of the lake is at B, finding an efficient control policy is more difficult. The trajectory coming from B in Fig. 8.5 has a null min-robustness value, since it leaves the viability set in 12 years. The control is already set to half the maximum allowed (–0.09/year), $\alpha = -6\%$ of the initial value $L(0)$ at B.

All these geometric indicators can also be used to define a particular action policy, for instance by maximizing the robustness as described in Sect. 8.4. The most important information concerns the choice of a particular control among all possible viable controls. For example, at point B, a lot of control values are still possible to keep the lake in a viable state. Generally a manager does not want to impose too strict regulation (for example, the maximum diminution allowed). In order to keep the robustness above a given threshold, the manager can anticipate and choose a viable control closer to the control available at the projection point $p(B)$. In the same way, if the manager wants to raise the robustness value of a lake at state A, he can propose a control close to the control available at the projection point $p(A)$.

In the next section we present the methods and tools that allow one to compute and use the distance map and projection to provide geometric information for a viability study.

8.3 Computing Geometric Robustness

The underlying concept of robustness is the distance to the boundary (of the significant sets of the viability study). It is used to define the robustness of states and trajectories, to describe the shape of the sets (with the largest maximal ball), to select appropriate control (through the projection point of a state). It is therefore essential to have at one's disposal an efficient method to compute the distance and projection. This is the subject of this section.

8.3.1 Practice of Geometric Method

In this section we develop the prerequisites that are often insufficiently taken into account when dealing with geometric methods. It is particularly important in the case of the study of viability since an efficient control policy depends on the result of the resilience and robustness study. Many choices are implicitly made before the computation of the distance, with impacts on the distance map. It is essential to underline the different options and their significance in order to really benefit from the geometric study. As for all geometric methods, the choice of the distance is a critical step of the geometric robustness study. A change of the distance can radically change the conclusion. For example, in Fig. 8.1a, a change from the Euclidean distance to:

$$d_2(x, y) = \sqrt{(3(x_1 - y_1))^2 + \left(\frac{1}{3}(x_2 - y_2)\right)^2}$$

transforms the right circle shape into the left ellipse. Both shapes share the same area, but the radius of the largest maximal ball in the case of the ellipse is three times smaller than in the case of the circle. This problem arises frequently with multiple criteria weighting.

However, the geometric study takes place after the building stage of the model, therefore this issue is generally already fixed when the distance map algorithm is applied.

In any case, a reflection about the distance is essential to interact with the human experts in charge of the model.

8.3.1.1 Nondimensionalized Variables

The viability kernel is a subset of $\mathbb{R}^n$, and the two main methods that are available to approximate the viability set generally perform homogenization and rescaling operation. So the distance map algorithm is generally applied to variables that are already nondimensionalized, and the distance induced by the inner product can be used directly. When it is not the case, model experts have to propose a suitable coordinate system in which the inner product is meaningful.

Two kinds of coordinate system transformation are basically useful, the Min–Max (MM) and the standard (s) transformation (see (8.1)) coordinate systems.

With the Min–Max transformation the model expert considers that a subset of the state space only has to be taken into account for the distance map transform. The range of every variable is reduced to the interval [0, 1] and the distance map applies to the unit hypercube. The main justification for the use of Min–Max transformation is that during the viability study, a constraint set in which the dynamical system should evolve is defined. This constraint set is a subset of the input state space. This set can be considered as the set of all possible initial states for the dynamical system, and then mapped to the unit hypercube for further use.

The Min–Max transformation is also particularly appropriate when the variables are in fact sensor measurements, since the accuracy of physical sensors is generally a function of the total range of values. For each attribute i the minimum and the maximum of the range are set to $Min(i)$ and $Max(i)$ and the new coordinate system is used to describe the state space.

The standard transformation is widely used in statistics to normalize sample data, using an estimate of the mean E_i and of the standard deviation s_i of each variable i. In the framework of viability study, $s(i)$ can be seen as a characteristic unit of measurement for variable i, and $E(i)$ a characteristic value for variable i. For example, $s(i)$ can be set to the range of the constraint set on variable i.

In practice, the choice of the transformation should be guided by the viability problem.

The new coordinate system is defined by (8.1).

$$y_i^{MM} = \frac{x_i - Min_i}{Max_i - Min_i} \quad \text{or} \quad y_i^s = \frac{x_i - E_i}{s_i}. \tag{8.1}$$

In this new coordinate system, the canonical inner product is meaningful.

8.3.1.2 Choice of a Distance

When the input state space has an inner product, the Euclidean distance which is derived from the inner product is generally used to compute the distance map unless some particular property is required. But it is not always the most appropriate. Other distances than the Euclidean distance can still be considered in order to compute geometric indicator for a viability kernel or a resilience basin. The use of a particular distance should be linked to the way the possible disturbances or uncertainties for each state variable can combine themselves.

$$\text{Euclidean distance} \quad d(x,y) = \|x-y\| = (\sum_i |x_i - y_i|^2)^{\frac{1}{2}} \tag{8.2}$$

$$L^1 \text{ distance} \quad d_1(x,y) = \|x-y\|_1 = \sum_i |x_i - y_i| \tag{8.3}$$

$$L^\infty \text{ distance} \quad d_\infty(x,y) = \|x-y\|_\infty = \sup_i |x_i - y_i| \tag{8.4}$$

With the Euclidean distance, the uncertainty (or error or possible perturbation) is distributed among all the variables: the amount of uncertainty sums over all the variables, following Pythagoras' theorem. The combined effect of a small disturbance a_i on every variable i is a larger disturbance a with size $\|a\| = (\sum_i |a_i|^2)^{\frac{1}{2}}$. With the L^1 (or Manhattan) distance, the combined effect of disturbances on different variables is simply their sum: $\|a\|_1 = \sum_i |a_i|$.

With the L^∞ (or sup norm) distance, the combination of perturbations of the same size on different variables does not change the size of the resulting perturbation: $\|a\|_\infty = \max_i |a_i|$. This means that every combination of disturbances are allowed, or, in other terms, the different sources of uncertainties or disturbances do not substitute for one another.

In the lake eutrophication problem, we used the Euclidean distance, since the two variables stand for a same chemical substance (coming from different sources). In other problems, with variables standing for different physical quantities, it can be preferable to use the sup norm distance. This is the case for example in the cheese ripening process (Mesmoudi et al. 2009), where state variables are rather different: temperature, mass, microorganisms respiration, etc.

8.3.2 *Distance Map and Projection Algorithm*

The method we use for computing the distance to the boundary was initially developed for classification systems (see Alvarez et al. 2010 for the general case and Alvarez 2004 for a detailed illustration on machine learning decision trees). It applies well to viability kernels and resilience basins which can be seen as classifiers. (States inside the viability kernel belong to the class *viable*, whereas states outside belong to the class ¬*viable*. States inside the resilience basin belong to the class *resilient*, whereas states outside belong to the class ¬*resilient*).

The distance and projection maps are computed with an adapted version of a discrete algorithm coming from mathematical morphology (Meijster et al. 2000). This algorithm is defined on a hyper rectangle of $\mathbb{N}^d$, and so the area of the state space for which the distance is computed has first to be mapped to a hyper rectangle in $\mathbb{N}^d$. This means in particular that two neighbours on any axis of the grid are at the same distance. The algorithm computes the exact distance of each point of the discretized space to a subset $\bar{S}$ of $\mathbb{N}^d$. We modified the original algorithm to compute the projection(s) onto $\bar{S}$. In our case, $\bar{S}$ is the set of non-viable or non-resilient states.

The algorithm 3 consists of two steps. The first step computes the distance and nearest point of the boundary on the first axis. It labels the points of the boundary if needed. The second step is called $d-1$ times and adds an axis to the previous subspace of dimension $k-1$. It updates the distance value and the list of projection points in the new subspace. To each norm is associated a function that specifies how the distance in dimension k subspace is computed from the distance in dimension $k-1$. For sake of simplicity, we consider that a unit hypercube of the input space is discretized onto a N points per axis d-dimensional grid G in $\mathbb{N}^d$.

For example, with the Euclidean distance, the square distance in dimension k subspace is given by a parabola: If $g_{k-1}(X)$ is the squared distance between X and some point of the boundary, computed on the first $k-1$ axes, and if u_k is the unit vector of axis k, then the squared distance in the dimension k subspace, $g_k(X)$ is at most $g_{k-1}(X)$. Applying the Pythagorean theorem, it is also at most $g_{k-1}(X - l.u_k) + (x_k - l)^2$ for $0 < l < N$, and the minimum value gives the result. But the computation and comparison of all these values is suboptimal.

The algorithm is optimal, since instead of computing all the distance values, it considers a set of distance functions and computes their lower envelope. For example, for the Euclidean distance, it considers the set F of parabolas:

$$\left\{F^X(i) = g_{k-1}(X) + (i - x_k)^2\right\}\ (0 \le x_k < N)$$

The square distance for all $x_k, 0 \le x_k < N$ is given by the lower envelope of F, that is, for each abscissa, the minimum value among all the values given by the different parabolas for this abscissa, as it can be seen in Fig. 8.6.

The key point is the fact that two parabolas in F intersect at most once in $\mathbb{N}^2$, and that the intersection is very easy to compute: F^X intersects F^Y when $2(x_k - y_k)$ divides $(x_k^2 - y_k^2 + g_{k-1}(X) - g_{k-1}(Y))$. Instead of storing all the parabolas values, only the intersection points and the vertices are stored.

This is the reason why it is optimal: The computation of the envelope is reduced to a matrix searching problem, whose complexity is of $O(N)$ in this totally monotonic case (see Hirata 1996 and Aggarwal et al. 1987 for details), so the overall complexity is in $O(N^d)$. The algorithm also works well for other distances, as long as this key point concerning the intersection of the building functions is maintained. The complementary set $\bar{S}$ of the viability kernel can be any subset of $\mathbb{N}^d$, so it contains $O(N^d)$ points. The complexity of algorithms which consider the distance of the different points of the state space to each of $\bar{S}$ points is therefore much higher.

For the Euclidean distance, the building function and the intersect function are defined by:

Algorithm 3 DistanceAndProjectionOnToSet. Sketch of the distance map algorithm.

Require: a map from $G = [0, N-1]^d$ in $\mathbb{Z}^d$ to $\{0, 1\}$. Points of the boundary are labeled by a function *label*$(x_1, ..., x_d)$. A building function $F(X)(i)$ and an intersect function according to the *norm*.

Ensure: Distance of each point of $[0, N-1]^d$ to $\bar{S}$ and the corresponding nearest point of $\bar{S}$

```
Procedure firstAxisDistance distance along the first axis
for all (x_2, ..., x_d) ∈ [0, N − 1]^{d−1} do
  for i ← 0 to N − 1 do
    d(i, l) = |i − l|
    A ← {d(i, l), 0 ≤ l < N and (l, x_2, ..., x_d) ∈ S̄}
    if A = ∅ then
      g(i, x_2, ..., x_d) ← ∞
    else
      j ← argmin{A}
      g(i, x_2, .., x_d) ← |i − j|
      p(i, x_2, .., x_d) ← label(j, x_2, .., x_d)
    end if
  end for
end for
return g, p
end procedure firstAxisDistance
if norm = Euclidean then
  g_1 ← g^2 post-processing for Euclidean distance
end if
for k ← 2 to d do
  Procedure AdditionalAxis example: Euclidean square distance
  for (x_1, ...x_{k−1}, x_{k+1}, ..., x_d) ∈ [0, N − 1]^{d−1} do
    for i ← 0 to N − 1 do
      recruitment of building functions
      A ← {F^{(x_1,..,l,...,x_d)}(i) = (d(i, l)^2 + g_{k−1}(x_1, .., l, .., x_d)}, with 0 ≤ l < N
      if min_{0≤l<N}{A} < ∞ then
        j ← argmin_{0≤l<N} A
        assignment following the building functions envelope
        g_k(x_1, .x_{k−1}, i, x_{k+1}, .x_d) ← F^{(x_1,..,j,...,x_d)}(i)
        p(x_1, .x_{k−1}, i, x_{k+1}, .x_d) ← p(x_1, .x_{k−1}, j, x_{k+1}, .x_d)
      end if
    end for
  end for
  end procedure AdditionalAxis
  if k = d then
    if norm = Euclidean then
      g_d ← √g_d
    end if
  end if
  return g_d, p
end for
```

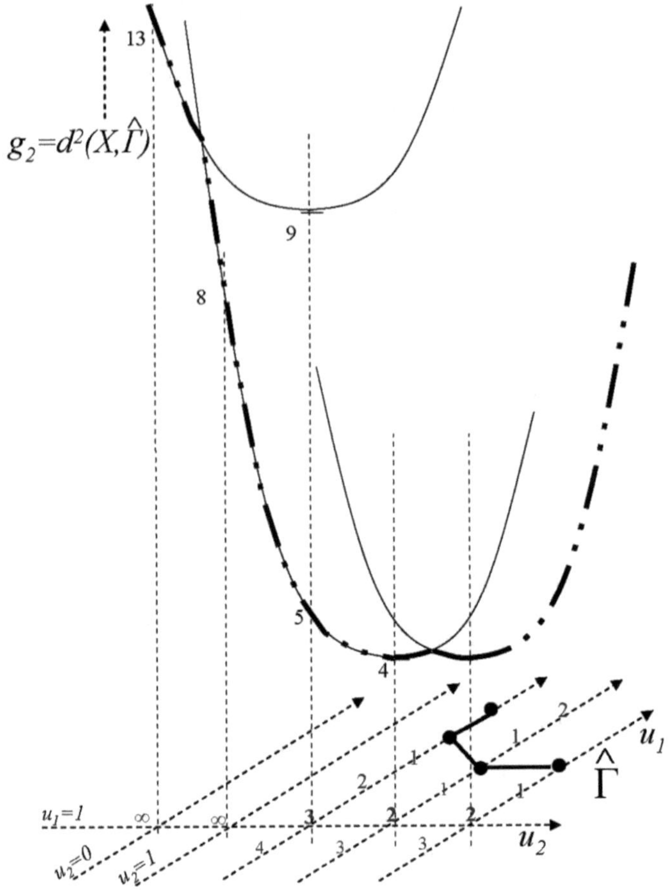

Fig. 8.6 Illustration of Algorithm 3. The distance along the first axis (in *grey*) is computed first. Then the parabolas are built along the second axis. Here, for $u_1 = 1$, along the second axis with $0 \leq u_2 \leq 3$, two parabolas are recruited with respective vertices $(2, 9)$ and $(3, 4)$. The *dot-dashed line* is their lower envelope. The second parabola is lower than the first one from point $u_2 = 1$, since the intersection is between 0 and 1. At the next step, $(u_2 = 4)$, another parabola will be recruited with vertex $(u_2 = 4, 2^2)$. Then the square distance will be assigned backwards, following the lower envelope: 4 at $u_2 = 4$, 4 at $u_2 = 3$, 5 at $u_2 = 2$, 8 at $u_2 = 1$ and 13 at $u_2 = 0$

$$F^X(i) = g_{k-1}(X) + (i - x_k)^2 \tag{8.5}$$

$$\text{intersect}(F^X, F^Y) = \text{truncate}(x_k^2 - y_k^2 + g_{k-1}(X) - g_{k-1}(Y)) \div 2(x_k - y_k). \tag{8.6}$$

For the sup norm, the distance in dimension k subspace is computed from the distance in dimension $k - 1$ with a truncated V-shaped function:

$$F^X(i) = \max(g_{k-1}(X), |i - x_k|) \tag{8.7}$$

The V-shaped function has its vertex at abscissa x_k and is truncated at the value $g_{k-1}(X)$, since the distance of the sup norm in dimension k is the max of the distance for the first $k-1$ axes and of the distance along axis k, that is $|i - x_k|$ at step i. The intersection of truncated V-shaped functions F^X and F^Y is given by the following formula (with $x_k \leq y_k$):

$$\text{intersect}(F^X, F^X) = \begin{cases} \max((x_k + g_{k-1}(Y)), ((x_k + y_k) \div 2)) \\ \qquad\qquad\qquad\qquad \text{if } g_{k-1}(X) \leq g_{k-1}(Y) \\ \min((y_k - g_{k-1}(X)), ((x_k + y_k) \div 2)) \text{ otherwise} \end{cases} \tag{8.8}$$

This algorithm is very efficient, since the complexity is in $O(d.N^d)$, which is optimal. It can also be parallelized on up to N^{d-1} processors. For example, the computation of the distance map for 10^9 points (20 points per axis in seven dimensions, or 1,000 points per axis in three dimensions) takes about 3 h on a 2.4 GHz processor (see Alvarez et al. 2010 for more details).

8.4 Robust Viability-Guided Management

We illustrate some viability and robustness guided policies using one of the language competition models, the bilinguals Minett–Wang model, presented in detail in Chap. 3.

8.4.1 Language Competition Model Description

In the bilinguals Minett–Wang model (Minett and Wang 2008), the population is made of three groups, the monolingual speakers of languages X and Y, and the bilingual speakers B. The model is two-dimensional, with the dimensions representing the proportions of X and Y speakers, on the x and y axis, respectively ($b = 1 - x - y$). The evolution of these two variables are governed by the following equations:

$$\begin{aligned} \frac{dx}{dt} &= (1-x-y)(1-y)^a s - xy^a(1-s) \\ \frac{dy}{dt} &= (1-x-y)(1-x)^a(1-s) - yx^a s \end{aligned} \tag{8.9}$$

where $s \in [0, 1]$ denotes the prestige of language X compare to the language Y one, and a is a parameter that models how the attractiveness scales with the proportion of speakers (for more details see Chap. 3).

The prestige measures the status associated with a language due to individual and social advantages related to the use of that language, being higher according to its presence in education, religion, administration and the media. We assume that public action can modify the prestige of a language, but that its variation at each time step is bounded:

$$\begin{aligned} \frac{ds}{dt} &= \quad u \\ u \quad &\in U := [-\bar{u}; \bar{u}]. \end{aligned} \tag{8.10}$$

The problem of maintaining a given level of monolingual speakers in both languages can be described by a subset of the state space, the constraint set, K, in the viability theory terminology:

$$K := [x_{min}; 1] \times [y_{min}; 1] \tag{8.11}$$

with $x_{min} > 0$ and $y_{min} > 0$.

It is possible to exhibit viability domains (see Bernard and Martin 2010) associated with the viability problem described by (8.9), (8.10) and (8.11). (A viability domain is a subset of the viability kernel, similar to it, but it is not maximal).

With this model, the variables are already normalized and the dot product is obvious. Since the boundary of the viability domain is computed directly (Bernard and Martin 2010), the distance algorithm computes an approximation of the distance to the boundary on a grid with no further conditions. Figure 8.7 shows a transparency view of the distance map inside the viability domain.

The distance map is then used to propose viable and robust policies.

8.4.2 Robustness to Perturbation and Uncertainty

Besides the viability kernel, viability theory gives at each state a list of controls that ensures that a state can stay in the viability kernel one step ahead. Nevertheless, all states in viability kernel are not equivalent, since a system in a viable state near the boundary can switch outside the viability kernel if subjected to unexpected perturbation. The distance map can help to define a more robust control policy than the standard viability control policy, which is based on the inertia-based avoidance control strategies. The same situation occurs in the resilience basin, and the distance map can help to take into account unexpected perturbations.

The following examples illustrate robust control policy in the case developed in Sect. 8.4.1. For simplicity we consider the viability domain, but the same approach would apply in a resilience basin (when several controls are available).

8.4.2.1 Inertia-Based Avoidance Control Strategies

Heavy Trajectories

Heavy trajectories (Aubin 1991) correspond to the choice at each time step of the control that minimizes the norm of the control rate of change. This means that the

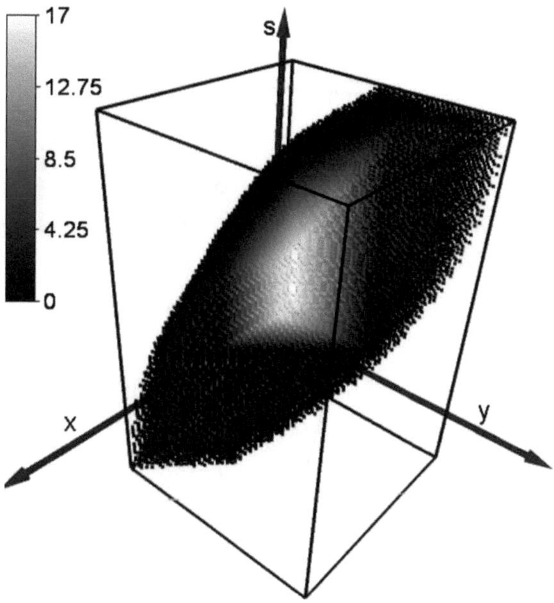

Fig. 8.7 Distance map of the language viability domain. A grid of 100 points per axis is mapped on the unit hypercube. Only the points of the viability domain are drawn. Their colour is a function of the Euclidean distance to the boundary. A *black bounding box* encloses the viability domain

control stays the same until it is necessary to change it to avoid leaving the viability kernel. Figure 8.8 shows an example of several heavy trajectories with random input state and control in the viability kernel. In general, with the control based on the Saint-Pierre (Saint-Pierre 1994) algorithm, these trajectories follow the flow of the dynamic system with constant control until they reach the boundary of the viability kernel. On the boundary of the viability kernel, when it does not coincide with the boundary of the constraint set, all the viable velocities belong to its tangent space. Consequently, the trajectories are stuck to the boundary until they encounter an area where the viability kernel coincides with the boundary of the constraint set. This is the reason why these trajectories stay a long time on the boundary, as shown in Fig. 8.9. These trajectories are different from the ones observed in the case studies of this book because the tool we used to compute the control actions prevents the system to get too close to the boundary (see Chap. 7).

Slow Trajectories

Slow Trajectories (Aubin 1991) correspond to the choice at each time step of the control with the smallest possible norm. This means that the system is not controlled except when it is necessary. Figure 8.10 shows a set of slow trajectories. In general these trajectories follow the flow until they reach the boundary of the viability kernel. As with heavy trajectories, they are then stuck to the boundary until they encounter an area where the viability kernel coincides with the boundary of the

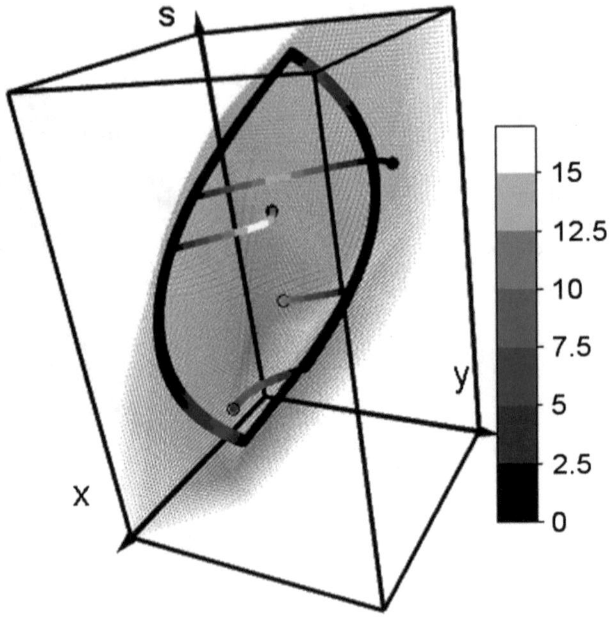

Fig. 8.8 State robustness of four heavy trajectories with random initial state (*black circles*). The colour of the points depends on the robustness value. The viability kernel is shown as a cloud of small points. The *black curved lines* correspond to trajectories on the boundary: Trajectories often stay on the boundary

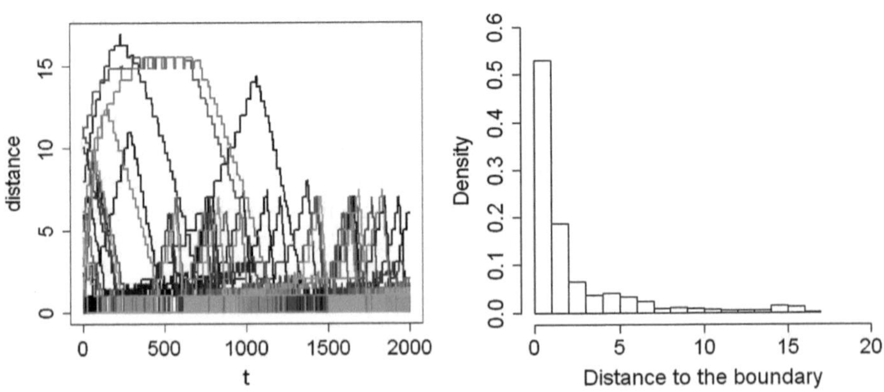

Fig. 8.9 Heavy trajectories for the language competition model. *Left*: Distance to the boundary as a function of time. *Right*: Mean frequency (density) of the distance to the boundary. Heavy trajectories stay on the boundary more than half the time

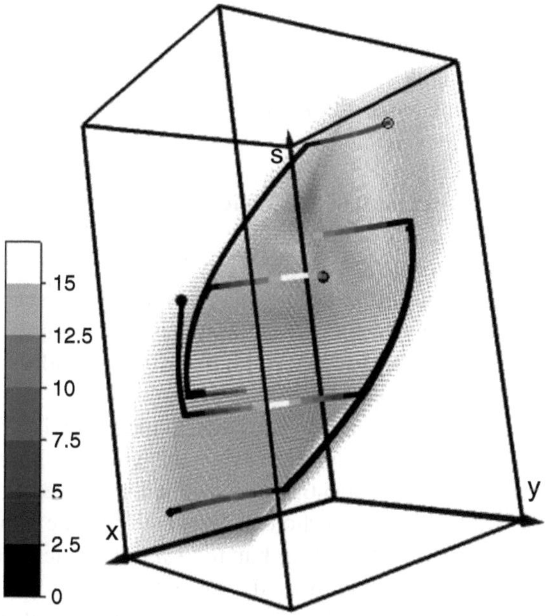

Fig. 8.10 State robustness of slow trajectories. *Black lines* are on the boundary of the viability domain: Slow trajectories stay very often on the boundary

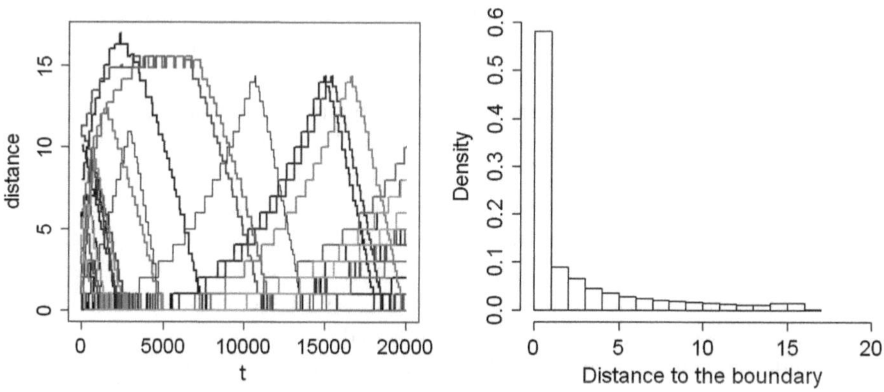

Fig. 8.11 Slow trajectories for the language competition model. *Left*: Distance to the boundary as a function of time. *Right*: Mean frequency (density) of the distance to the boundary

constraint set. Slow trajectories, like heavy trajectories, stay a lot of time on the boundary, as shown in Fig. 8.11.

Standard control policies lead to trajectories that can stay for a long time very close to the boundary of the viability kernel. This problem cannot be easily solved by strengthening the constraints: being inside the viability kernel of a smaller constraint set ensures that, for given perturbation strengths, the system state remains inside the initial constraint set, but not in the viability kernel. So, without a complete study of the resilience basin of the new viability kernel, there is no guarantee that a trajectory, even in the more constrained viability kernel, will resist unexpected perturbation.

8.4.2.2 Geometric-Based Control Strategies

The main objective of a geometric-based strategy is to propose control policies that are robust to unexpected perturbation or uncertainties in the state variables. The same concern is addressed in Chap. 7. The principle is then not only to follow viable (or resilient) strategies but also to remain far from the boundary of the viability kernel (or resilience basin) if possible.

For this purpose we design a strategy we call "Avoidance with Threshold". This implies a modification of the heavy strategy in the viability kernel. The control remains constant until the state robustness comes below a given threshold (for the heavy strategy, the value of the threshold is zero). When the robustness reaches the threshold, the control is still viable, but for sake of anticipation, the control corresponding to the projection point of the current state on the viability kernel boundary is applied. The idea behind this heuristic is that the control that applies on the boundary takes the flow into account.

A variant of this method is applied in Chap. 7, selecting the control that optimizes the value of the SVM several steps ahead.

The robustness-based strategy, in order to be efficient, needs a fast and reliable way to access to the distance and projection on the boundary. This is provided by the distance map algorithm described in Sect. 8.3.2.

Figure 8.12 displays the distance to the boundary of the viability kernel as a function of time for three trajectories that start at the same initial point but follow avoidance strategies with different threshold values. It shows that the avoidance strategy is very efficient: The heavy strategy (avoidance strategy with threshold equal to 0) governs a trajectory whose distance to the boundary of the viability kernel is often equal to 0. The avoidance strategies with strictly positive thresholds govern trajectories whose distance to the boundary of the viability kernel never crosses the threshold once this threshold is reached.

To underline the efficiency of the avoidance strategy, we perform different numerical experiments. We choose randomly different initial points and compute the trajectories governed by avoidance strategies with different thresholds including zero. For each trajectory, we compute the distance to the boundary of the viability kernel as a function of time and then the mean relative frequencies of this distance over all trajectories for each different strategy. Figure 8.13 shows the results computed over 24 trajectories. Most of the time, the trajectories are above the distance threshold. The frequency of the distance values smaller than the threshold are not null because the randomly chosen initial point may have a distance to the boundary smaller than the threshold. But once the distance to the boundary passes

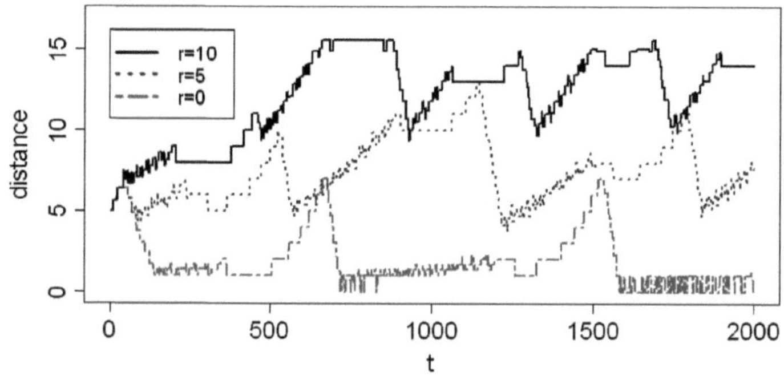

Fig. 8.12 Trajectories from the same initial state point with different robustness threshold values (r). The avoidance with threshold strategy keeps the system away from the boundary

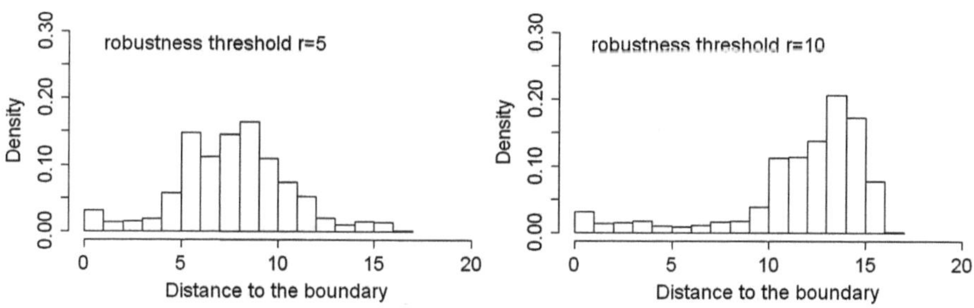

Fig. 8.13 Mean frequency of the distance to the boundary with different robustness threshold values

over the threshold it never returns below it. The avoidance strategy with threshold is very effective for the language competition model.

Finally, we go back to the trajectory robustness definitions proposed in Sect. 8.2.3, and we compute the different robustness values of trajectories starting at the same point but following avoidance strategies with different thresholds. Figure 8.14 shows the results for different starting points. Except for the min-robustness, where both values can be equal to 0, the robustness of the trajectories with threshold is always strictly above the corresponding robustness of the standard heavy trajectory.

Remarks

It is worth noting that the same approach can be followed for slow trajectories.

Geometric information can also be used to propose directly a control that aims at maintaining the system far from the decision boundary. For example, in the language competition model, it is possible to propose a control directed towards the centre of the largest maximal ball, as shown in Fig. 8.15. This type of control changes

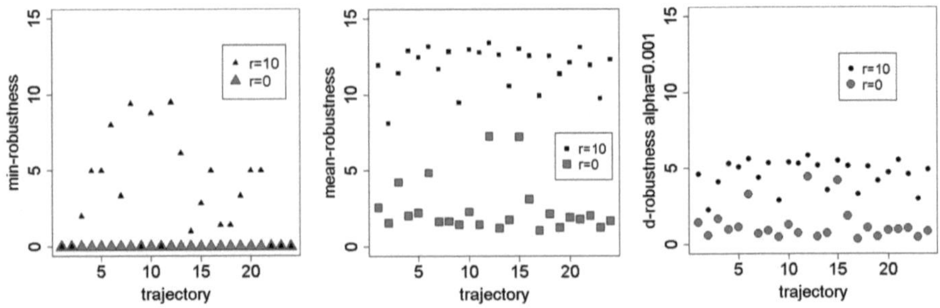

Fig. 8.14 Comparative robustness of trajectories stemming from the same input point, with and without distance threshold

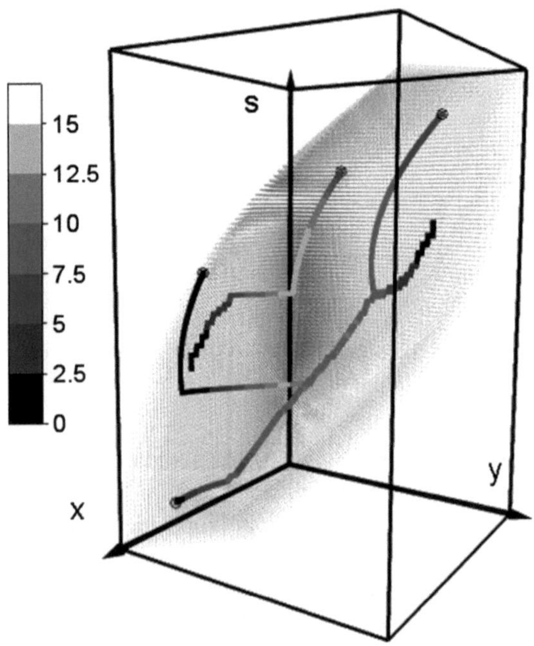

Fig. 8.15 Trajectories with control directed towards the centre of the largest maximal ball. Points of the viability kernel are drawn as *small dots*. Trajectories are *thick lines*. Colour is a function of the distance to the boundary

radically the pattern of the trajectories, which can stay much longer in the white area (the farthest area from the boundary). However, this strategy can be unsuccessful depending on the flow. Some trajectories have a low robustness value despite the attempt to pull the trajectory towards the centre of the largest maximal ball. They cannot evolve anymore and stay in a low robustness state.

8.5 Conclusion

The definition of the viability kernel ensures the existence of an action policy that keeps the state of the system inside the constraint set. Analogously, for states inside the resilience basin, there exists at least one policy that allows the system to reach the viability kernel in finite time. However, what happens if a perturbation occurs and causes a jump of a given length of the system state? Does it remain in the viability kernel or in the resilience basin? The response to these questions is given by the distance to the boundary of these sets.

To compute an approximation of this distance on a regular grid, we have used a distance transform algorithm. This information associates each state belonging to the viability kernel or the resilience basin with its distance to the boundary and gives information about the robustness of this system state to perturbations. Besides, geometric robustness of the states inside the viability kernel or resilience basin can also be used to propose several definitions of trajectory robustness, which can be used in turn as new optimization criteria.

Heavy or slow control policies minimize at each time step the control norm or rate of change (which can be seen as control costs). These standard control policies lead to strategies lacking in robustness: trajectories can stay for a long time very close to the boundary of the viability kernel.

Geometric information, taking into account the distance to the set boundary, can be proposed to adapt the standard control strategy, in order to maintain the system away from the viability kernel or resilience basin boundary. We have proposed a geometric-based strategy: when the state robustness comes below a threshold, the control corresponding to the projection point of the current state is applied, even if the present control is still viable (or resilient). To implement such a strategy, we have modified the distance transform algorithm to include the approximation of the projection onto the boundary. And we have shown that it does indeed lead to more robust trajectories in the language competition model.

Other strategies could also be proposed, taking into account more local geometric information, such as the distance to the skeleton. In the case of a resilience basin, the robustness could be taken into account as well as the resilience value, to select the appropriate control to drive the system back to the viability kernel. (The geometric robustness could be taken into account even more directly in the definition of the cost function that is used to define the resilience value).

Appendix: Mapping the Exact or Approximate Viability Kernel or Resilience Basin onto a Discrete Grid

The DistanceAndProjectionOnToSet algorithm (3) provides the exact distance map to a discrete subset $\bar{S}$ of $\mathbb{N}^d$. It is relevant to use it to compute an approximation of the distance to a viability kernel or a resilience basin with some general hypotheses.

Depending on the method that is used to define explicitly the viability kernel or the resilience basin, an approximation of its boundary can be computed. In the other cases, an approximation of the set itself is computed by the viability or the SVM approximation algorithm. The approximation of the distance is not the same in this case.

Case Where the Boundary Is Known

When the subset $\bar{S}$ of $\mathbb{N}^d$ is an approximation $\hat{\Gamma}$ of the true boundary Γ, the algorithm computes directly the exact distance of the points of the grid $G = [0, N-1]^d$ to $\hat{\Gamma}$.

For simplicity, we assume that Γ is included in the unit hypercube H of E. We define G_N the regular grid with N points per dimension: G_N has N^d elements $\hat{x} = (k_1, \ldots, k_d)$, $k_i \in \{0, N-1\}$ corresponding to $x = (k_1/(N-1), \ldots, k_d/(N-1))$ in E. The discretized boundary $\hat{\Gamma}_N \subset G_N$ is defined as follows:
Let $\hat{x} \in G_N$ and let x be its corresponding point in E,

$$\hat{x} \in \hat{\Gamma}_N \text{ if and only if } d(x, \Gamma) \leq \frac{\sqrt{d}}{N-1} \tag{8.12}$$

($\sqrt{d}$ is the diagonal length of the unit hypercube and $(N-1)$ the rescaling coefficient).

Theorem 8.1. *Let $y \in H$ and $\hat{x}_N$, the nearest point of y in G_N. Then the exact distance of $\hat{x}_N$ to the discretized boundary $\hat{\Gamma}_N$, approximates the distance of y to the viability kernel boundary Γ in H, $\Gamma \cap H$ as N goes to infinity.*

Proof. Let $P(y)$ be the projection of $y \in H$ onto $\Gamma \cap H$ and $\hat{P}(\hat{x}_N)$ be the projection of $\hat{x}_N \in G_N$ onto $\hat{\Gamma}_N$. For the Euclidean distance (but similar inequalities exist for other distances), let $\epsilon_N = \frac{\sqrt{d}}{N-1}$. By construction of $\hat{\Gamma}_N$, there exists $\hat{P}(y) \in \hat{\Gamma}_N$ such that $d(\hat{P}(y), P(y)) \leq \epsilon_N$ and $P(\hat{x}_N) \in \Gamma \cap H$ such that $d(\hat{P}(\hat{x}_N), P(\hat{x}_N)) \leq \epsilon_N$. Then, thanks to triangular inequality,

$$\begin{aligned} d(y, \Gamma \cap H) &\leq d(y, P(\hat{x}_N)) \\ &\leq d(y, \hat{x}_N) + d(\hat{x}_N, \hat{P}(\hat{x}_N)) + d(\hat{P}(\hat{x}_N), P(\hat{x}_N)) \\ &\leq d(\hat{x}_N, \hat{P}(\hat{x}_N)) + 2\epsilon_N \end{aligned}$$

For the same reason, $d(\hat{x}_N, \hat{P}(\hat{x}_N)) \leq d(y, \Gamma \cap H) + 2\epsilon_N$ □

When we consider the Euclidean distance, it is also possible to approximate the Euclidean projection onto the viability kernel boundary.

Theorem 8.2. *Let $y \in H$ and $\hat{x}_N$, the nearest point of y in G_N. Then the projection of $\hat{x}_N$ onto the discretized boundary $\hat{\Gamma}_N$, approximates the Euclidean projection* *onto the viability kernel boundary in H, $\Gamma \cap H$, outside the skeleton*[1] *of the connected parts of the class areas, as N goes to infinity.*

Proof. Let $P(y)$ be the projection of $y \in H$ onto $\Gamma \cap H$ and $\hat{P}(\hat{x}_N)$ be the projection of $\hat{x}_N \in G_N$ onto $\hat{\Gamma}_N$. $(\hat{P}(\hat{x}_N))_N$ is a bounded sequence, so it has a convergent subsequence toward a point Q. Since $d(\hat{P}(\hat{x}_N), \Gamma) \leq \frac{\sqrt{d}}{N-1}$, $Q \in \Gamma$. Moreover, from Theorem 8.1, $d(y, Q) = d(y, P(y))$. Since y doesn't belong to the skeleton, $P(y) = Q$. □

Case Where Only the Viability Set Is Known

In this case the viability algorithm or the SVM approximation computes an approximation S of the viability set. The boundary of the viability set Γ is not explicitly defined. Nevertheless, instead of defining the discretized boundary, we can defined the discretized complementary set of S, $\hat{\bar{S}}$ in G:

$$\hat{x} \in \hat{\bar{S}}_N \text{ if and only if } d(x, \bar{S}) \leq \frac{\sqrt{d}}{N-1} \tag{8.13}$$

Then, for the same reason as above,

Theorem 8.3. *Let $y \in H$ and $\hat{x}_N$, the nearest point of y in G_N. Then the exact distance of $\hat{x}_N$ to the discretized set $\hat{\bar{S}}_N$, approximates the distance of y to the viability kernel boundary ∂S in H, $\partial S \cap H$ as N goes to infinity.*

References

Aggarwal A, Klowe M, Moran S, Shor P, Wilber R (1987) Geometric applications of a matrix-searching algorithm. Algorithmica 2:195–208

Alvarez I (2004) Sensitivity analysis of the result in binary decision trees. In: Proceedings of the 15th European conference on machine learning, Lecture notes in artificial intelligence, vol 3201. Springer, Heidelberg, pp 51–62

Alvarez I, Martin S, Mesmoudi S (2010) Describing the result of a classifier to the end-user: geometric-based sensitivity. In: Proceedings of the 22d European conference on artificial intelligence, IOS Press, Amsterdam, pp 835–840

Aubin JP (1991) Viability theory. Birkhauser, Basel.

Bernard C, Martin S (2010) Building strategies to ensure language coexistence in presence of bilingualism. Cemagref. Internal Report.

[1]The skeleton S of A is the set of the centres of all maximal open balls in A (see Matheron 1988 for a study of its topological properties. We suppose here that the skeleton is the complementary set of a dense open set.)

Carpenter SR, Ludwig D, Brock WA (1999) Management of eutrophication for lakes subject to potentially irreversible change. Ecol Appl 9:751–771
Hirata T (1996) A unified linear-time algorithm for computing distance maps. Inform Process Lett 58(3):129–133
Martin S (2004) The cost of restoration as a way of defining resilience: a viability approach applied to a model of lake eutrophication. Ecology and Society 9(2): 8. http://www.ecologyandsociety.org/vol9/iss2/art8/
Matheron G (1988) Examples of topological properties of skeletons. Image analysis and mathematical morphology, Academic Press, London, pp 217–257
Meijster A, Roerdink J, Hesselink WH (2000) A general algorithm for computing distance transforms in linear time. Morphology and its applications to image and signal processing, pp 331–340
Mesmoudi S, Alvarez I, Martin S, Sicard M, Wuillemin P-H (2009) Geometric analysis of a capture basin: application to cheese ripening process. In: European conference on complex system, October 21–25, 2009, Warwick, UK.
Minett J, Wang W (2008) Modelling endangered languages: the effects of bilinguism and social structure. Lingua 118:19–45
Saint-Pierre P (1994) Approximation of the viability kernel. Appl Math Optim 29:187–209
Saltelli A, Chan K, Scott M (2000) Sensitivity Analysis. Wiley, San Francisco
Serra J (1988) Image Analysis and Mathematical Morphology. Academic Press, London

Index

G. Deffuant and N. Gilbert (eds.), *Viability and Resilience of Complex Systems*,
Understanding Complex Systems, DOI 10.1007/978-3-642-20423-4,

MIX
Papier aus verantwortungsvollen Quellen
Paper from responsible sources
FSC® C105338

If you have any concerns about our products,
you can contact us on
ProductSafety@springernature.com

In case Publisher is established outside the EU,
the EU authorized representative is:
Springer Nature Customer Service Center GmbH
Europaplatz 3, 69115 Heidelberg, Germany

Printed by Libri Plureos GmbH
in Hamburg, Germany